Quick Changeover
Simplified

Quick Changeover Simplified

The Manager's Guide to Increasing Profits with SMED

By Fletcher Birmingham
and Jim Jelinek

CRC Press
Taylor & Francis Group
Boca Raton London New York

CRC Press is an imprint of the
Taylor & Francis Group, an **informa** business

A PRODUCTIVITY PRESS BOOK

CRC Press
Taylor & Francis Group
6000 Broken Sound Parkway NW, Suite 300
Boca Raton, FL 33487-2742

© 2007 Taylor & Francis Group

CRC Press is an imprint of Taylor & Francis Group, an informa business

No claim to original U.S. Government works

ISBN: 9781563273490 (pbk)

Visit the Taylor & Francis Web site at
http://www.taylorandfrancis.com

and the CRC Press Web site at
http://www.crcpress.com

TABLE OF CONTENTS

CONTENTS

GETTING STARTED

Let's get right to the point. Setup reduction is a concept that involves three basic components:

- Simplification
- Effective communication (mostly listening)
- Extreme follow through with a passion for process discipline

These components are the cornerstones of setup reduction. The binding ingredient is action—that is, you have to actually *do it* to get the benefit. The concepts and proven practices are straightforward and easily learned. Effective implementation is critical! The biggest problem most people face is not getting started. The second biggest problem is poor or no follow through with process discipline. These mistakes can be fatal.

The purpose of this book is to help those companies or employees, who *understand the need* and *have the desire* to implement a quick setup and changeover program, *but aren't sure how to proceed*. We want to show you how easy it is to get started and how effective the program can be. If you already have a program, we will give you tips on how to make it even more effective. Our ultimate goal is to help you create an environment in which everyone understands and is an active participant in reducing setup and changeover times. Setup reduction must be a way of life—not just another program. Embrace this concept and your company will become better and stronger. The bar of excellence is continually rising at a faster rate. Meeting this challenge should be *fun*.

The main takeaway value from this book is the simple, effective methods to get you started. Once you get started and begin to see the payback involved, you'll be hooked and on your way to the ultimate goal of creating a culture of lean thinking.

A second and equally important benefit is the realization that lean thinking alone is not enough to survive in today's globally competitive environment. Innovation, agility in effectively responding to changing markets, and successfully finding, training, motivating, and retaining the best people will be equally important.

The third takeaway is the knowledge that while you can make great strides using internal resources, it won't be enough. To constantly innovate and improve your company, you also need to go outside your walls to gather, understand, and then to implement great ideas, methods, and technology. Don't limit yourself to homegrown solutions. Wal-Mart founder Sam Walton spent much of his time studying the competition and implementing and improving those ideas that worked. Send your people (including your best machinists—not just management) to tool shows, other manufacturing plants, tooling vendors, etc., to learn about what's new and what works. Then implement and improve on those ideas.

Finally, ask yourself this question: Are you radical enough in your rate of improvement? Consider this scenario: In the past two years, raw material prices have risen 20 percent to 30 percent. Annual insurance increases of 20 percent are common. Wage increases of 3 percent to 3.5 percent are normal. On top of these increases, customers and foreign competition are putting severe downward pressure on prices. Nevertheless, given these examples, we believe that incremental improvements of 5 percent to 10 percent are not enough. We all need to be thinking in terms of 25 percent to 30 percent improvement just to survive. Are you achieving these rates of improvement?

Meeting the Global Marketplace Challenge

All companies today compete in a global marketplace. Companies in the United States have several disadvantages when competing in a global economy. These challenges include:

- High labor costs, including costs associated with pensions, insurance, etc.
- Shortages of young people entering and being trained in manufacturing

- Increasing sophistication and productivity of foreign competitors
- Technology that can make your products obsolete

The combination of these and other factors can give countries such as China, India, Brazil, Vietnam, and Romania huge cost advantages. Companies that don't continually improve productivity, lower costs, and improve quality will find themselves losing to these competitors.

Another factor facing U.S. manufacturers is the growing move toward commoditization on both a national and global basis. Customers today have access to more information than ever before. Information about and assistance in buying products and services across the country or across the world is only a keystroke away on the Internet. This trend will only continue to grow.

Although all companies like to think that their products and services are special, you need to ask, "How special is my product or service *in the eyes of the customer*? If my product or service is special, will the customer be willing to pay a premium? Or, is there a similar product that's perceived as good enough or substantially equal at a lower price?" Positioning your company as the low-cost producer will provide a defense against such perceptions and will, in all likelihood, give you a competitive edge if consistently offered.

A quick setup program is a critical component in the overall strategy of implementing lean manufacturing to obtain operational excellence and low cost. In other words, quick setups are crucial if you want to be a low-cost, high-quality producer. Without quick setups, you will not achieve this status—and without this status, you will not remain competitive.

Operational excellence and low cost alone are not enough. They are just prerequisites. Today's most successful companies, such as Dell and Wal-Mart, realize that they need to be customer focused in the extreme. If you want your company to not only survive but also thrive in the future, you need to make sure that every activity benefits the customer. Activities that don't benefit the customer must be eliminated.

Why You Need a Quick Setup and Changeover Program

Companies are implementing setup reduction programs for many reasons. Foremost is *economic survival*; in other words, to stay ahead of the competition. Increased profitability, job security, stability, and growth are right up there, as well as wanting to be the best at what they do.

Have you ever heard your people say:

- "But we have gotten along okay so far."
- "We are doing well."
- "We are better than most."
- "Everything will be okay."
- "Listen, we are the best already. We can't do better."
- "We don't have the resources."
- "We are just taking care of the customer as best we can."
- "We tried that before, but it didn't work."

If you are hearing statements like these, your company is in trouble. They indicate a failure to see what level of excellence is currently required and the level of excellence that will be required in the future.

> *"Even if you're on the right track, you'll get run over if you just sit there."*

In his book *Good to Great*, Jim Collins correctly states that "Good is the enemy of great."[1] Similarly, believing you are already good enough is the enemy of truly being good enough to survive much less achieving greatness. It's best to stay hungry and never be satisfied. That way, you stay committed to constant improvement.

---

1. Jim Collins, *Good to Great: Why Some Companies Make the Leap . . . and Others Don't* (New York: HarperCollins, 2001), 1.

It is no longer enough to compare yourself to where you were five years ago and be satisfied with incremental progress. Today, you must benchmark yourself against the best in the world. As former GE CEO Jack Welch says, "You need to look at reality as it is, not as you wish it to be."[2]

---

2. Jack Welch and John A. Byrne, *Jack: Straight From the Gut* (Lebanon, IN: Warner Books, 2001).

CHAPTER 1

Committing to Quick Changeover

The Changeover Race Is Won or Lost in the Pits

Imagine that you are at a NASCAR race. Your favorite driver is running on fumes, so he has to make a pit stop for some gas and a new set of tires. As he pulls into the pit, his pit crew slowly gets up and surrounds the car. After they stare at the car and talk among themselves for a minute or two, the pit crew chief finally says, "How about if Bob and Bill change the tires this time, while Jeff and I pour the gas and wash the windshield. Then, next time, we'll switch—Bob and Bill can pour the gas and clean the windshield, while Jeff and I change the tires." After they all shake their heads in agreement, they then start rummaging through a pile of tools to find the ones they need. Meanwhile, the driver sits and waits, watching his chances at being a winner diminish with each passing car—knowing that he'll be bringing up the rear because he can never make up the time lost in the pits.

You will never see this scene at a NASCAR race, but it occurs every day in the business world. Many manufacturers lose valuable time when operators change over from one job to the next. Quick setups and changeovers are vital to remaining competitive in today's business world. Like a NASCAR pit crew, manufacturers should strive to change from one product type to another with the least amount of downtime possible.

How SMED Gives Your Company a Competitive Edge

The concept of reducing setup and changeover times isn't new. It originated from the work of Dr. Shigeo Shingo, who successfully reduced the setup and changeover time for large presses in the Toyota Production System. Dr. Shingo had to increase production capacity without purchasing new equipment. First, he studied machines to find ways to make them run faster. While his discoveries were often helpful, he wasn't satisfied with what he learned. One day, he returned to his office early and discovered a machine sitting idle. Much to his surprise, he realized that all of the atten-

> ### What's In It for Your Company?
>
> A setup-reduction program will
> - ◆ Simplify your manufacturing processes
> - ◆ Improve the quality of your products
> - ◆ Increase throughput
> - ◆ Permit smaller lots
> - ◆ Make your company more competitive
> - ◆ Save jobs

tion paid to that machine's cycle time was to no avail. After a production order was completed, the machine lay idle while workers leisurely gathered the appropriate tooling and material for the next order. At that moment, Dr. Shingo realized that reducing setup and changeover time is critical to achieving full production capacity, and began to focus on reducing setup time. During one of his studies, he stated that changing production equipment from the last good piece to the first good piece should take less than 10 minutes. Hence, the term "Single Minute Exchange of Dies" (SMED) was born.

SMED is now often referred to as "quick setup and changeover" or "setup and changeover reduction." For simplicity's sake, we'll refer to this concept as "quick setup" or "setup reduction." Although SMED was developed to improve die-press and machine-tool setups, the principles apply to changeovers for all types of product setups. Reaching the single-minute range might not be possible for all setups, but a quick setup program can dramatically reduce setup time in almost all cases.

A quick setup and changeover program will:

- **Simplify your manufacturing processes.** A quick setup and changeover program simplifies processes and makes manufacturing jobs easier and more fulfilling for employees. Happier employees can lead to a lower employee-turnover rate.

- **Improve the quality of products.** A quick setup and changeover program improves the quality of your products by helping you better define, simplify, and control your manufacturing processes.

- **Increase throughput.** A quick setup program allows an increase in throughput, which improves deliveries. Improved deliveries will help your customers sell more products to their patrons—and your customers, in turn, will need to purchase more products from you.

- **Permit smaller lots.** Manufacturers often produce goods in large lots because the long setup time makes it too costly to change processes frequently. Producing large lots for this reason has several disadvantages, including:
    - *Inventory waste.* Storing what isn't sold costs money and ties up company resources without adding value.
    - *Quality loss.* Storing unsold inventory increases the chance that it will have to be scrapped or reworked.
    - *Delay waste.* Customers must wait for the company to produce entire lots rather than the quantity they need.
    - *Nonstandardized setups.* Infrequent setups often aren't standardized; thus, they are difficult and risky.

- **Make your company more competitive.** A quick setup and changeover program reduces the time, cost, and resources associated with switching from one manufacturing job to the next. The program can also reduce costs, such as inventory holding costs and the cost of lost time due to delays. You can pass these savings on to the customer, making your company more competitive.

- **Save jobs.** Not implementing a quick setup program makes your company noncompetitive because it needs to absorb the cost of lost potential savings that could have benefited your company and customers. Ultimately, jobs will be lost. By reducing setup times, you can remain competitive and save those jobs.

Many companies already have excellent programs in place. Some companies, however, haven't started such a program. This unfortunate circumstance might be the result of:

- Shortsightedness
- Lack of belief in the benefits to be gained

- A belief that they're too busy to implement the program

- Exposure to similar programs that failed to deliver in the past

- A feeling that they're already the best and can't get better

In addition, there is a common misconception that setup reduction costs money. Don't be fooled into believing it. Quick setup and changeover make you money! These companies don't realize how effective a quick setup program can be and how easy it is to start one. These companies don't realize that the cost of not taking action is high (i.e., the payback philosophy), that setup reduction saves jobs, and that the benefits will be shared by all—the company, the employees, and the customers.

> **SETUP REDUCTION DO'S AND DON'TS**
>
> **Do**
>
> ✔ Make the decision to reduce the time your organization is spending on setup operations. NOW is the best time.
>
> ✔ Understand that your decision must be backed up by passion and commitment. Successful implementation demands follow through.
>
> ✔ Realize that you must take the time to invest in your program and your people.
>
> **Don't**
>
> ⊗ Wait or hesitate.
>
> ⊗ Get lulled into thinking that operational excellence and low cost alone are enough to survive in today's marketplace. They are just prerequisites. Your organization also needs to be customer focused to the extreme—and must do it now!

> *"Change starts when someone sees the next step."*
>
> —William Drayton

Don't let your company fall in with this crowd. Someone at your company has to see and take the first step, and that someone must be you. It doesn't matter whether you're the company owner, general manager, plant manager, lead man, machinist, controller, or support staff member. It must be *you*. Remember the basic building blocks—simplicity, communication (mostly listening), and passionate follow through.

11

In this section, we will look at some examples of how to calculate the financial benefits.

The True Cost Is in *Not* Implementing Lean Thinking: Payback Analysis

Quick setup and changeover make you money! To implement a quick setup and changeover program, you initially take only those actions that have a payback. The payback period is typically less than one year, unless you take actions that involve a major capital expenditure. Therefore, the cost of not implementing a quick setup and changeover program is the savings that don't accrue after the payback point is reached—the point at which the savings from an improved setup and changeover process has paid for the one-time cost of implementing the activity that led to the improvement. After the payback point is reached, the river of savings doesn't suddenly stop. Instead, it continues to flow year after year. By not reducing setup and changeover times, these potential savings are never realized. Note: In these and similar calculations, we reduced estimated savings 30 percent to 50 percent, to be conservative. We also estimated costs on the high side for the same reason.

Now, let's consider the following examples:

Example 1: New Tool Presetter

- Cost $16,000
- Savings estimated at 12 hours per week at $30 per hour for 52 weeks, or $18,720 per year
- Payback period $16,000 ÷ $18,720 × 12 months = 10.25 months

In other words, after 10.25 months, you'll receive savings at the rate of $18,720 per year net of any costs that have been repaid. If you didn't acquire the tool presetter, you would lose these savings, which is the cost of not having done a setup reduction program. In a sense, these savings become an annuity or future stream of earnings.

Example 2: New Portable Hardness Tester

- Cost $7,500
- Savings estimated at 8 hours per week at $30 per hour for 52 weeks, or $12,480 per year
- Payback period $7,500 ÷ $12,420 × 12 months = 7.2 months

After 7.2 months, you get a net savings/annuity of $12,480 per year.

Think of these savings in terms of time, which can be used to produce more products with the same number of people. Our simple estimates do not take into consideration the added intangible benefit of faster throughput, which allows us to produce products faster and improve delivery time.

Payback analyses can be integrated into a company's processes. For example, at Flo-Tork, an Ohio-based manufacturer of pneumatic and hydraulic rack and pinion rotary actuators, any proposed improvement involving a purchase that has a one-year (or less) payback is authorized immediately. If a proposed improvement falls within the two-year payback period, the company spends some additional time researching the change to verify the accuracy of the costs and savings estimates. When the estimates are verified, it authorizes the change. If a proposal falls beyond the two-year payback period, the company does a more extensive study of costs and savings. In addition, it

---

Financial Benefits

Consider the following savings that these companies enjoyed:

◆ A steel mill reduced the setup time for rolling and drawing steel wire from an average of 3.5 hours to an average of 90 minutes, which led to increased mill utilization by 11 percent and an annual cost savings of $133,000.

◆ A manufacturer of pigment and chemical dispersions for thermoset plastics reduced the setup time for forming pellets from 83 minutes to 43 minutes, which led to an increase in production capacity by 140,000 pounds per month and a monthly savings of more than $100,000.

factors in the intangible benefits, such as increased quality, enhanced customer service, and improved throughput.

Test Drive the Mini Setup- and Changeover-Reduction Program

An ounce of effective action is worth a ton of theoretical discussion. You can take several simple steps to prove to yourself that setup reduction really works. After you implement these steps—which we call the *mini setup- and changeover-reduction program* because it's a trial-sized version of the real deal—and reap the benefits, you'll no longer question the need to implement a comprehensive program.

> *"There comes a moment when you have to stop revving up the car and shove it into gear."*
>
> —David Mahoney

The mini setup- and changeover-reduction program involves getting ideas from several employees over coffee and doughnuts one Saturday morning. However, you can implement any number of variations of this program. You can meet on a weekday morning instead of a Saturday or get together for lunch rather than breakfast. You can implement the mini program in any manufacturing area you feel would work best. If you own the company or you are a top-level manager, you might want to start in a critical area in which setups take a long time; any reductions in that area will be beneficial and easy to spot. Alternatively, you might want to choose a less critical, off-line area just to see whether the mini program works. If you are a machinist or lead man, you might want to start in your own area and challenge your coworkers to find improvements.

Figure 1-1 outlines the steps in the mini setup- and changeover-reduction program. This program came about because I (Jim) was skeptical, just as you may be. As the president and chief executive officer of Flo-Tork, I truly believed that while the company had an excellent manufacturing facility, it could do more and in fact *needed* to do more. In retrospect, I never realized how much more improvement was possible. We needed to ensure that we

Saturday Morning Coffee and Doughnuts

Requirements

Pad of paper and a pen
Coffee and doughnuts (or possibly bagels for the health conscious)
Belief, enthusiasm, and determination
Small investment of time
Willingness to listen
Willingness to provide immediate and complete follow through
Willingness to celebrate success with all involved

How It Works

Setup reduction sponsor or enthusiast (you) approves (if in an appropriate position) or obtains approval (i.e., ask management) to do the following:

1. Pick an area in which setup is a major component of the job.

2. Discuss with the operators the benefits of setup reduction and explain that you are trying a pilot program to determine whether there is anything to this "stuff." Tell them you would like to have them shut the machine down for 2½ hours on Friday so that they can answer the questions in step 3, then meet with you Saturday morning to review their answers.

3. On Friday afternoon, have the operators shut down the machine for 2½ hours. Have them answer, as a group, the following questions:
   - What are the major problems you face each day?
   - What should management be doing that it isn't doing now to support you?
   - What would you recommend management do to make the setup easier?
   - What types of tooling, fixtures, handling equipment, procedures, and so on would be helpful?
   - What can support people do to make your job easier?
   - What more can your area do if management helps you?

4. On Saturday morning, meet with the operators to discuss their answers. Turn these answers (and any other recommendations) into action items in a to-do list. The action items need to be

(continued on next page)

Figure 1-1. Mini Setup-Reduction Program

specific. For each item, the group needs to designate a person to be in charge (however, the entire group is responsible for the success of the venture), determine a timeframe for completion, and perform a simple payback calculation. The list needs to be published and progress reported until the list is completed.

5. On Monday, you must act immediately on one or two items that the group agrees will have a high payback. Action must be highly visible and effective. The item does not have to be large in scale. It could be a small action, such as acquiring a tool. Belief in the program is achieved by people seeing evidence of commitment and tangible, positive results. Some of the items on the to-do list might take longer to act on, but it's imperative that each item be acted on. Even if the group decides later to drop or modify an item, it must be resolved.

6. Celebrate and reward the group when all the action items are implemented.

7. Document the setup procedures and policies you have implemented and enforce process discipline.

8. Repeat this process area by area.

Figure 1-1. (*Continued*)

maintained the level of operational excellence necessary to respond to the relentless onslaught of global competitors.

I had read a lot about lean manufacturing and had attended many lean manufacturing workshops. Setup reduction was of particular interest to me because Flo-Tork mainly does small-lot production, but I wasn't ready to start a full-blown program until I had more confidence that the program would deliver as promised. I decided to test the waters with what I now call the mini setup-reduction program.

I went to my two lead men in the machining center area, which is an area in which setups involve a major time investment. I explained why Flo-Tork needed to reduce setup times; I basically told them what I wanted to do and why (you'll read more about the reasons in Chapter 2). Then I explained that for 2½ hours on Friday afternoon, I wanted them to shut down the machine and collaborate on what the company could do to improve the setup processes in their area. I told them that on Saturday morning, I would buy coffee and doughnuts and that we—the plant manager, the two lead men, and I—would discuss their ideas. I also told them that I would truly listen to what they had to say and follow through.

We met that Saturday morning. The two lead men had filled several pages of legal-sized paper with ideas. They had so many simple yet effective ideas that we actually decided to get started that day. Here are some of the things they brought to our attention:

- They wanted time to clean and reorganize the cabinets and work areas in each machining center. The cabinets and work areas were filled with accumulated junk and clutter, which meant that the operators often spent time hunting for tools. By removing the junk, cleaning the area, and reorganizing the tools, the operators could quickly find what they needed. In reality, they simply wanted time to implement the 5Ss—sort, set in place, shine, standardize, and sustain—another lean manufacturing concept.

- They recommended that each machining center have its own cart that contained tie-down bolts and clamps. At that time, there was only one cart for all three machining centers, which meant that the operators sometimes had to hunt for the cart or, worse, use a makeshift solution if some of the bolts or clamps they needed were in use.

- They recommended that we buy a small quantity of commonly used tool holders and boring bars for the machines to avoid a situation in which an operator didn't have the proper tools.

- They asked us to ensure that the tool puller be trained to do the job more completely and thoroughly.

- They asked for better lighting in certain areas so that they could better see what they were doing.

On Monday, the machines in each machining center weren't started until the cabinets and work areas were uncluttered, cleaned, and reorganized. Depending on the machines, this took between four to six hours. New carts with complete sets of tie-down bolts and clamps arrived in the machining centers within a week, as did the recommended tool holders and boring bars. By the end of six weeks, all of the lead men's ideas were implemented.

At the end of the first week, we were all surprised at how much progress we had made in so little time and with so little investment. In addition, the lead men were quite pleased that we had taken the time to listen to them and to make their jobs better. Their attitude, as well as the attitude of the whole shop, was more upbeat because they saw that management was serious.

We started to see the results of the mini setup reduction program almost immediately. Besides seeing a positive change in the employees' attitudes, we found that we were, in fact, able to complete the setups in significantly less time. Employees from other areas came to me to ask if we could do the same process in their area. The culture was starting to take root.

The main lesson I learned was that the key to progress is to take the time to listen to employees and follow through. I realized that the potential for improvement was real. Based on that realization and the mini setup-reduction summary, which Figure 1-2 shows, I decided to go forward with a full-scale program.

You can use the mini setup- and changeover-reduction program again and again until you are ready to implement a full-scale program. The full-scale setup- and changeover-reduction program involves seven simple steps:

1. Prepare the workforce.
2. Select the project and assemble the team.
3. Create a baseline for the project.
4. Separate internal from external.
5. Convert internal to external.
6. Streamline efforts.
7. Standardize improvements and celebrate successes.

Costs

Time

Friday afternoon question & answer session: 2 operators × 2.5 hours
= 5 hours

Saturday morning meeting: 4 people (2 operators + 2 managers) ×
4 hours = 16 hours

Organize 3 areas × 6 hours each area = 18 hours

Train tool puller better = 4 hours

Lighting enhancements = 2 hours

Materials

Coffee and doughnuts = $15

Cart's tie down bolts = $500

Redundant boring bars = $800

Lights = $500

Costs Summary

Time: 45 hours at $30 per hour = $1,350

Material: $1,815

TOTAL COST: $3,165

Estimated Savings

Savings: 3 people in affected areas at 2 hours per week (total of
6 hours per week) at $30 per hour for 52 weeks = $9,360

Payback period: $3,165 ÷ $9,360 × 12 = 4.1 months

Figure 1-2. Mini Setup-Reduction Program Summary

Be the Champion

Someone has to champion and start the quick setup and changeover program at your company. That someone must be *you*.

> *"There are no secrets to success. It is the result of preparation, hard work, and learning from failure."*
>
> —Colin Powell

It doesn't matter whether you are the company owner, general manager, plant manager, lead man, machinist, controller, or a support staff member. It must be you. Make the decision to give the mini setup- and changeover-reduction program a try. Don't be fooled into believing that setup reduction costs money. Use the examples in this chapter to calculate a return on investment before buying new tools or fixtures. Remember the basic premise—simplicity, communication (mostly listening), and follow through.

POINTS TO REMEMBER

- ☛ Add passion to all your endeavors. Very little is accomplished without it.
- ☛ Make your setup-reduction program fun!
- ☛ Make it a challenge.
- ☛ Make it a dream.
- ☛ Make it an adventure.

CHAPTER 2

Selecting the Project and the Group

OK, you've decided to champion the setup-reduction program—congratulations on a wise and courageous decision! As the champion, you might be wondering what to do next. As Figure 2-1 shows, you have several important responsibilities including picking an area in which to start the program and selecting someone to fill the role of group leader and/or assume a facilitative role. You might be the best person to fill the group leader and facilitative roles, so your first task is to determine how active you want to be in the setup-reduction program.

The champion of the setup-reduction program needs to

◆ Pick an area in which to start the program

◆ Select a group leader

◆ Select someone to assume a facilitative role

◆ Demonstrate commitment and support for the program through actions as well as words

◆ Make sure that all successes are rewarded

Figure 2-1. The Champion's Responsibilities

Determine Your Level of Involvement

Active involvement can take several forms and still be effective. As the champion, you can be totally immersed in the minute-by-minute details or you can set the parameters and make sure that the group has the resources it needs to succeed. The advantages of being totally immersed in the program include the following:

- You can offer a broader business perspective to the group if needed.
- You can offer technical expertise if needed.
- You can help facilitate the setup-reduction program.
- You can visibly demonstrate your commitment to the program.

The disadvantage of being actively involved is that, if you are a manager or the company owner, your staff might not feel comfortable speaking their minds or talking about the areas in which improvement is needed when you are around.

One alternative is for you to give the group time to conduct its studies and make its recommendations. The group can then discuss its recommendations with you. At this point, you can assist in the payback studies and make sure that all approved improvements are properly funded to ensure timely completion.

Regardless of the level of your involvement, it is absolutely critical that you show true commitment to the program through actions as well as words. You must make sure that *all* successes are rewarded.

Selecting the Project Area and the Group Leader

The decision about the area in which to implement the first setup-reduction project and the decision about who should lead the group are not mutually exclusive. It is best if the group leader is a supervisor or manager in the area you want to improve. The reasoning is that this person has easy access to the resources and information the group needs. In addition, the group leader typically selects the specific setup process in that area to improve. Thus, you need to keep in mind an area's potential for success as well as whether a manager or supervisor in that area will enthusiastically embrace the setup-reduction program.

Things to Consider When Selecting the Project Area

If you successfully implemented a mini setup-reduction program, that area might be a good place to start the larger program. If you didn't implement a mini program, follow the guidelines we used to select an area for the mini program.

You might be tempted to work on the simplest setup operation because it won't hurt production too much when the machines are taken off-line. Conversely, you might want to tackle the most complicated setup operation because it's been troublesome. However, neither setup operation is the best candidate for a first project. Improving a simple setup operation doesn't offer much payback. Trying to improve the most complicated setup operation might lead to frustration and poor results because the process is too complex for a newly formed group that is just learning how to reduce setup time.

The best approach is to choose a run-of-the-mill setup operation that produces a common item. The operation should also be typical in what it needs in terms of tooling, supplies, and people. If the group works on a run-of-the-mill setup operation, the group has a good chance of achieving good results. The most critical requirement, though, is to start somewhere.

Things to Consider When Selecting the Group Leader

The group leader is critical to the success of the setup reduction group. You need to choose a manager or supervisor who you think will most effectively carry out the responsibilities. Figure 2-2 outlines those responsibilities.

The group leader

◆ Decides which setup-reduction operation to tackle first

◆ Decides which individuals should be in the group

◆ Takes care of meeting scheduling and logistics

◆ Makes sure that all the group's suggestions are implemented

---

Figure 2-2. The Group Leader's Responsibilities

As Figure 2-2 shows, one of those responsibilities is to decide who should be in the setup-reduction group. The group should include between six and twelve members, in addition to the group leader. Ideally, the group members should be volunteers. However, getting volunteers for the first setup-reduction group might be difficult. In that case, the group leader should ask the appropriate employees to join the group. The group should represent a cross-functional, cross-work area mix—you need cross-pollination to bring in new ideas and challenge the status quo.

SETUP REDUCTION DO'S AND DON'TS

Do

✔ Demonstrate commitment and support for the setup-reduction program through actions as well as words.

✔ Focus on one area and follow through until tangible results are received.

✔ Showcase the results to others.

✔ Have the group leader be a supervisor or manager of the area you want to improve.

Don't

⊗ Pick projects from a hat.

⊗ Start on a project that's too simple or too complex.

⊗ Let the group leader dictate what the group should do.

To make sure that the group includes a mix of employees from different functions and work areas, the group leader should follow the ⅓-⅓-⅓ formula. One-third of the group should consist of the operators who normally run the job—they are the experts. However, sometimes they are at a loss for ideas because they are too close to the existing methods. Therefore, the next third should consist of people who have skills similar to the first third but work on a different product line or in a different production area. These group members will offer fresh ideas because they work on a different operation. The last third should be people in administrative and support functions, such as design engineering, process engineering, quality assurance (QA) engineering, maintenance, accounting, or human resources. The group doesn't need to include someone from every administrative or support department, but a few of those departments should be represented. These group members are helpful because they can change policies and procedures that might inhibit quick setup. Also, they gain a better understanding of how their work impacts production areas.

Another responsibility of the group leader is to take care of the group's meeting scheduling and logistics. For example, the group leader needs to determine when it would be a good time for the group to spend one to three days working on a quick setup program. Because the group leader is a manager or supervisor, he or she can look at the production schedule to see when there is an upcoming slowdown in jobs. The group leader needs to notify everyone involved about when and where to meet and take care of any logistics, such as reserving a meeting room and getting the necessary supplies (e.g., a flipchart).

During the group meetings, the group leader might take on the role of facilitator, which means that he or she needs to make sure that the group is following the setup-reduction process. It is imperative that everyone participates and follows the agreed upon ground rules. The group leader can offer suggestions during group meetings, but he or she must not introduce his or her preplanned solutions. The group leader

WARNING
The facilitator must not
◆ Take over the group
◆ Impose his or her preplanned solutions on the group
◆ Shut down debate

can inadvertently destroy the energy of a group by starting off with comments, such as "This is what we'll do . . . ." or "I did a little advance work on this and figured that we can improve by . . . ."

Finally, it's the group leader's responsibility to keep a written record of the group's suggestions for improvement. To keep track of the suggestions that haven't been implemented, the group leader should create a punch list, as explained in Chapter 8.

The Critical Role of Facilitator

The setup-reduction program requires that someone take on a facilitative role. This choice is critical to the success of the program. The person who takes on this role can be the group leader, the company owner, a manager, any staff member (e.g., machinist, operator), or an outside consultant. In other words, anyone can take on the facilitative role (i.e., act as a facilitator), as long as that person possesses certain attributes, outlined in Figure 2-3.

The facilitator's role is to make sure that the members learn the principles and techniques for reducing setup time. The facilitator

The facilitator must

◆ Believe in the process

◆ Have a passion for follow through

◆ Have a passion for process discipline

◆ Be enthusiastic and be able to instill enthusiasm among the group members

◆ Have excellent communication skills, especially good listening skills

◆ Have good people skills, including the ability to bring out the best in people and make the best use of various personalities so that every group member contributes

◆ Uphold the process's basic ground rules (which are covered in detail in Chapter 3)

◆ Keep the group members going in a positive direction and make sure that they don't get lost in the "But we've always done it this way" or "It won't work" mind-set

◆ Serve as a link between the group members and management

◆ Remain open minded

◆ Share credit for successes with the group

Figure 2-3. The Facilitator's Critical Attributes

also needs to make sure that everybody is participating. If a group member isn't contributing, the facilitator must bring that person into the process so that he or she becomes an active participant. If group members are becoming defensive and getting into arguments, the facilitator must get the group back on track.

The facilitator should not simply write an idea on a chalkboard or flipchart and say, "Let's do it," which is a risk you take when managers or other people in authority facilitate groups. They are so accustomed to leading that it's hard for them to extricate themselves from that role.

Although facilitators shouldn't force their ideas on the group, they can help guide the group. Often, facilitators can present an idea in the form of a question. That way, the group still ultimately makes, and therefore buys into, the decision.

The facilitator does not have to be directly responsible for the results, even though he or she should feel a strong sense of ownership for improving the setup operation. In small organizations, you might choose someone who is in the project area as the facilitator. In that case, the person will need some guidance to make sure that he or she can stay in the role of facilitator and not fall back into the role he or she typically fills. That person has to understand that the facilitator's role is about creating the right environment for the setup-reduction program and about making sure that the group follows the process. The one thing that the facilitator must *not* do is impose his or her own will on the group.

POINTS TO REMEMBER

- ☞ Pick a project that's likely to be successful
- ☞ Pick a product that will be around for awhile so you can reap the savings from future setup and changeover
- ☞ Choose a project leader who has good people skills
- ☞ Choose employees who offer their ideas and suggestions

CHAPTER 3

Preparing the Group

Once the setup area is selected and the members notified, it's time to hold the first meeting, which will set the tone for the setup-reduction program. In this meeting, you, as the program champion, need to:

- Introduce the program and explain why it is so important to the company and to the employees.

- Give an overview of how the setup-reduction process works and discuss management's commitment to this process and the group.

- Discuss everyone's expectations, set the ground rules, and explain a little bit about group dynamics and paradigms.

Introducing the Program

If you simply tell the group that they are going to implement a quick setup program without explaining what the program means to the company and to them, most employees will immediately become defensive. Their first impression will be, "Do you mean that we're not working fast enough? After all, the words 'quick setup' imply that we must work faster." Other members might object because they fear that going faster will lead to unsafe practices. That is why you need to explain upfront what quick setup is and why your company needs to implement a setup-reduction program. You need to stress early and often the importance of safety, and that you are interested only in the *safest* setup operations. You also need to let the employees know what's in it for them so that they will feel a sense of ownership in the outcome and see the benefits that result from their participation.

> *"Do You Mean That We're Not Working Fast Enough?"*
> —Anonymous

What Quick Setup Is and Why the Company Needs It

How do you impart knowledge about the quick setup program among the group members? How do you help them realize the critical role that reducing setup time plays in the overall strategy to grow the business and protect their careers?

Perhaps the best way is to simply get the group together and explain in general terms that they're going to learn new methods for reducing setup times.

The quick setup program isn't about working harder; it's about working smarter.

When you talk about the program's purpose, focus on the employee's interest in maintaining and creating jobs by becoming more efficient than competitors. Be careful of the language you choose. While cutting costs may sound beneficial; many people equate that with cutting jobs.

One effective approach is to verbalize *as well as* document on a flipchart the program's purpose. For example, you might say and write, "The goal is to reduce setup time on extruder No. 3." Immediately after, however, you need to assure the group members that there is nothing wrong with the way they are currently performing their jobs. For example, you might say, "Although we're here to focus on reducing setup on extruder No. 3, we're not saying that there is something wrong with the way we do it now. We're just saying that is it necessary for us to jointly look for ways to *improve* the current setup procedures. We will not criticize anyone for the past. Everyone has complete amnesty."

During this discussion, one question that usually pops up is, "*Why do we have to improve?*" (If there are managers present at the meeting, many employees won't feel comfortable enough to ask this question aloud, but it's likely running through their minds.) One way to answer that question, whether or not it's asked aloud, is to present a sheet of frequently asked questions (FAQs) about setup-reduction programs. Figure 3-1 is a sample FAQs sheet. As you can see, the FAQs sheet is short and to the point. These FAQs are generic enough that you can simply incorporate them in a handout to give to your group. Alternatively, you can customize them so that they are specific to your company.

Q: Why do we need to improve?

A: Today, companies must compete in the global marketplace. Companies in the United States have some disadvantages when competing in the global economy. The disadvantages include high costs for health care and other types of insurance, FICA contributions, unfavorable currency exchange rates, pension contributions, and high legal and labor costs. If our company doesn't continue to improve, the cost structures of U.S. companies relative to those companies in China, India, and other nations will make it hard for our company to compete. Therefore, we need to implement a quick setup program to remain competitive.

Q: How can a setup reduction program help us improve?

A: By finding ways to make the setup process more efficient, we can reduce the amount of time that machines are idle between production runs. As a result, we can move more products through the facility, which means that our company can produce more products, and therefore we can handle more orders from customers. The increase in productivity means that our organization is stronger.

By finding ways to eliminate waste, our company can save money and can pass those savings along to customers. We want to pass those savings to customers so that they can lower their costs and can increase their business, which, in turn, means buying more products from our company.

Q: What exactly do you mean by "quick setup"?

A: Quick setup is a way to reduce machine downtime. Specifically, it's the amount of time taken to change a machine from good product to good product in clock time, not labor time. In other words, it's the amount of time it takes to tear down an old setup on a machine, get the new setup in place, and get the machine up and running again, making quality products. It can't be just any product; it has to be a quality product that meets specifications. At the same time, the setup procedure needs to be done in a way that's safe and sustainable.

Figure 3-1. FAQs About the Setup-Reduction Program

What Employees Have to Gain from Quick Setup

To make sure that the members have a stake in the quick setup program, you need to tell them what's in it for them. The answer to this question is important for employees at all levels, from the president, CEO, or plant manager to those on the shopfloor. Figure 3-2 will help you answer this question. You can use Figure 3-2 as a handout or as a script for a presentation. You can use it as is or adapt it to meet your company's needs.

A setup-reduction program will

◆ *Make your job easier.*
The key is not to work harder but to work smarter by streamlining the process. We want to take out unnecessary steps and provide you with the best tools, fixtures, documented procedures, inspection reports, and anything else you need. We can remove wasted motion, job stress, and frustration. Setup reduction simplifies, and simpler is always easier.

◆ *Make your job more fun.*
When done properly, quick setup programs are fun. You will have the opportunity to interact with other employees in a positive environment of creativity, sharing, and learning. It's exciting to watch a group grow together forming a common bond of working toward goals that benefit everyone. You'll learn more about yourself and other employees, and you'll be exposed to new ideas. There is great pride and fulfillment in seeing your group's efforts result in an improved process. You'll also share the rewards with the group, which is always fun.

◆ *Make your job better.*
We need you to help us make your job better. Everyone wants to do his or her job as efficiently as possible. No doubt, there have been times when you have said, "I can show them how to do it better if they would only listen." The quick setup program offers you the opportunity to have a say in how your job is performed. We want your ideas. We will truly listen to those ideas and follow through. We

(continued on next page)

Figure 3-2. What's In It for the Employees?

will be pleased when someone shows us a way to do something better. By getting a group together, we can optimize the process. Together, as a group, we can ask the right questions, generate ideas, and take advantage of everyone's experience.

◆ *Protect your job by improving quality.*
The process of simplification, which inherently improves planning, organization, and execution, always improves quality. First-piece setup scrap is usually greatly reduced. In addition, the time lost to reworking scrapped parts is greatly reduced or eliminated because the simplified procedures make sure that perfect parts are manufactured. The customer receives high-quality parts on time (no delays), which helps us delight our customers and sell more products.

◆ *Protect your job by allowing us to fulfill customer orders faster.*
Faster delivery gives our customers a competitive edge, which will enable them to sell more to their patrons and, in turn, buy more from us.

◆ *Protect your job by making us the low-cost, high-quality producer.*
This program will allow us to improve our methods and eliminate waste from our manufacturing processes. Savings will result from higher productivity, less downtime, the ability to reduce inventory waste through efficiently producing smaller lots, and increased flexibility to meet changing customer demands.

◆ *Protect your job by making us more competitive in a global environment.*
Today, companies must compete in a global economy. Other countries, such as China and India, have tremendous advantages in terms of lower wages and lower labor costs (e.g., costs associated with insurance, pensions, and legal issues, which make our goods more expensive). These competitors are becoming increasingly sophisticated and are quite formidable. Most important, they are hungry and willing to do what is needed to obtain and retain business. Only by a constant effort to improve our efficiency, utilize technology, and be totally customer driven can we compete in today's world. It's not enough just to sell our product.

Figure 3-2. (*Continued*)

Laying the Foundation for the Setup Reduction Program

Our experience shows that it takes between two to three hours to lay the proper foundation for starting a setup reduction program. Here's a brief summary of topics to address and a guideline for how much time to spend on each topic:

Overview and management commitment	20 to 30 minutes
Expectations and concerns	20 to 30 minutes
Ground rules	20 to 30 minutes
Group dynamics	20 to 30 minutes
Challenging the way you do things	20 to 30 minutes
Break time	20 to 30 minutes

You have to be careful not to let any of these topics get out of hand. If you find yourself spending more than 30 minutes on one of these topics, it is likely you are getting bogged down on tangents. We have seen a lot of folks get into trouble in this area. Conversely, you need to avoid the temptation to rush through these topics. Group members shouldn't feel they are being pushed through, but at the same time, they shouldn't be falling asleep.

Providing an Overview and Demonstrating Management's Commitment (20 to 30 minutes)

When you get together with the group for the first time, it's important that you not only provide an overview of the setup reduction process but also demonstrate *your* commitment to it.

The overview. When you outline the setup reduction process, explain to the group that they will be gathering information about a setup operation by observing and documenting and, in some cases, videotaping (this step is not needed for the mini setup, and will be discussed in detail in Chapter 4) that operation. Let them know that to videotape a setup session, you must shut down the machine for a day or more.

When you share this detail, group members often have mixed reactions. Some are astounded that management would shut down a machine just to study a process. Others are anxious about whether the company can afford to shut down the machine because of the lost production time. Let the group know that the company can't afford *not* to shut down the machine. A good analogy you might want to share is the story of the lumberjack:

> A lumberjack is working long and hard to cut down a tree. An onlooker sees the lumberjack struggling, so he goes over to the man and asks, "Why is it taking so long to cut down the tree?"
>
> "Well, I think the blade is dull," replies the lumberjack.
>
> "Why are you using a dull blade?"
>
> "I don't have time to sharpen it."

The difference between "involvement" and "commitment" is like an eggs-and-ham breakfast: The chicken was "involved"—the pig was "committed."

—Unknown

That's an old story, but its message still rings true. Sometimes you have to push aside day-to-day production to make time for improvements. If you are unwilling to make the commitment to shut down a machine, you're cheating the company out of some real positive payback.

After addressing any concerns that the group members might have about shutting down a machine, let them know that they will then:

- Analyze the information they videotaped and documented so that they can create a baseline for the current setup practices.
- Put together ideas for improvement.
- Implement their improvements.
- Time how long the new setup operation takes.
- Compare that measurement to the baseline.
- Standardize the new setup procedure if an improvement occurs.

- Document the new procedure and instill a passion for process discipline.
- Provide training on those new procedures to the other operators.
- Celebrate their successes and recognize their achievements.

Proving management's commitment. During the first meeting, you need to express your commitment to the setup-reduction process and to the group. Let them know that you are not going to put the burden on them and then walk away. Prove to them that you are committed to listening to what they say and to following through on their recommendations. Demonstrate *by your actions* that you are committed to recognizing, sharing, and celebrating their successes.

That is a key point: Your actions *must* match your rhetoric. One important way to show your commitment is to set aside time for the group's improvement efforts. You can't simply tell the group to think about how to reduce setup time and to come to the meetings with their ideas. If they work eight hours a day trying to meet production quotas, when are they supposed to think about how to reduce setup time—at home in the shower?

Even if a group member thinks of an improvement at home or during the commute to work, one person alone can't flesh out and test the improvement. It takes a group effort to look at what is involved in a setup operation. A setup operation isn't only about the operator and the machine. It is about everything and everyone that affects the setup operation, ranging from the tools and the tool puller to first-article inspection and the quality assurance (QA) inspector. That is why making improvements must be a group effort among the operators, management, and the support staff—and you have to give the group time to work together and focus on improvements instead of day-to-day production.

One of the benefits of the setup reduction process lies in the fact that the group interaction lays the foundation for systematic thinking. All related functional areas that affect the setup operation will be represented as part of the system and not as just an isolated area.

Discussing Expectations and Concerns (20 to 30 minutes)

The next item on the agenda is a discussion of the expectations and concerns of everyone involved in the setup-reduction program. This seemingly miniscule activity helps create an atmosphere of mutual respect.

You need to ask every meeting participant two questions: "What are your expectations as a member of this group?" and "What are your concerns about what we are working on?" Then write each person's expectations and concerns on a flipchart. Some people will choose not to say anything—that's okay for the moment. However, you must later go back to those people and ask them again. Getting feedback from every group member helps engage them in the process and helps them develop a feeling of ownership for the quick setup program. Figure 3-3 shows some commonly held expectations and Figure 3-4 lists some commonly expressed concerns.

◆ To reduce the setup time for the operation

◆ To maintain a safe work environment

◆ To improve or maintain quality levels

◆ To create a punch list for items that must be done at a later date

◆ To complete all items on the list

Figure 3.3. Common Expectations

◆ What's wrong with the way we do it now?

◆ Another program of the month

◆ Management will never let us do this

◆ We don't have time for this—we have parts to make

Figure 3.4. Common Concerns

You must be careful not to be a wordsmith and edit what the group members say. Make it a point to use their language. The words to which you should pay particular attention are the words that are unique to your company, your culture, and the group. You need to understand and use those words as precisely as possible. This might mean that you need to ask a group member, "Can you please explain what you mean by that?"

After you have asked all of them about their expectations and concerns, you should ask everyone to review the flipchart and see whether they have anything else to add. If no one does, close the discussion by saying that they'll review this list at the end of the day and periodically thereafter to determine whether their expectations are being met and their concerns are being addressed. The list of expectations and concerns will serve as a kind of a checkpoint to make sure that the group is on track.

Setting the Ground Rules (20 to 30 minutes)

For people to successfully work in a group together, they need ground rules. As the facilitator, you have to make a short list of them, using the rules in Figure 3-5 as a guide. Then you should ask the group members if they want to add any.

The ground rule of creating a blameless environment heads the list because it's crucial for a successful setup-reduction program. You need to make it clear that the group must focus on the future, not on the past, and that criticizing what has happened in the past and finger-pointing are not going to be allowed. Members need to understand that they haven't been brought together to say that the operators have been doing the setup operation wrong all along. Nor have they been brought together to say that the operators haven't been doing a good job. They have been brought together because today's environment is tough and the company needs to be tougher to be competitive. They have been brought together to look for ways to reduce setup times, thereby making the company a tougher competitor.

The ground rules can come in handy later when unwanted situations arise. For example, if the members get stuck in the mind-set that they can't do anything differently, you can turn to the ground rules on the flipchart and say, "Let me ask you, are we really following the ground rule of keeping an open mind?" If the group

◆ Create a blameless environment—no finger-pointing.

◆ Everyone participates.

◆ Keep an open mind.

◆ Keep a positive attitude.

◆ Everyone has an equal voice.

◆ Truly listen to what other people are saying.

◆ Ask questions because there's no such thing as a dumb question.

◆ Understand that the question "Why do we do this?" isn't a criticism—it's just a question.

◆ Never sit in silent disagreement.

◆ Treat others as you want to be treated.

◆ Check ranks at the door.

◆ Create a bias for action—just do it!

◆ Learn together.

◆ Have fun.

Figure 3-5. Common Ground Rules

gets mired in a conflict, you can point to the "Treat others as you want to be treated" ground rule and ask, "Are we really following this ground rule?"

Explaining Group Dynamics (20 to 30 minutes)

We have found that you need to explain group dynamics. If you don't, the group will become lost and frustrated when conflict emerges, which inevitably happens during the setup-reduction program. Explain that the group will go through four stages of evolution: forming, storming, norming, and performing.[1]

Forming. Forming is the stage in which each participant makes the transition from individual worker to group member. When a group is first forming, the members might feel proud of being chosen yet, at the same time, wonder how much they will be able to

---

1. Peter R. Scholtes, Brian L. Joiner, Barbara Streibel, and Dale Mann, *The Team Handbook*, 2d ed (Madison, WI: Oriel, 1996).

contribute. Typically, the group members are quiet because they are trying to get accustomed to one another and they are not quite sure what's expected of them. A bit later in the forming stage, some group members might cautiously test the boundaries of acceptable group behavior and test the group leader's guidance both formally and informally.

Because there is so much going on at once, progress in meeting the group's objective is very slow during the forming stage, as Figure 3-6 illustrates. During this stage, the group members should focus on establishing ground rules and developing an action plan to fulfill the group's objective.

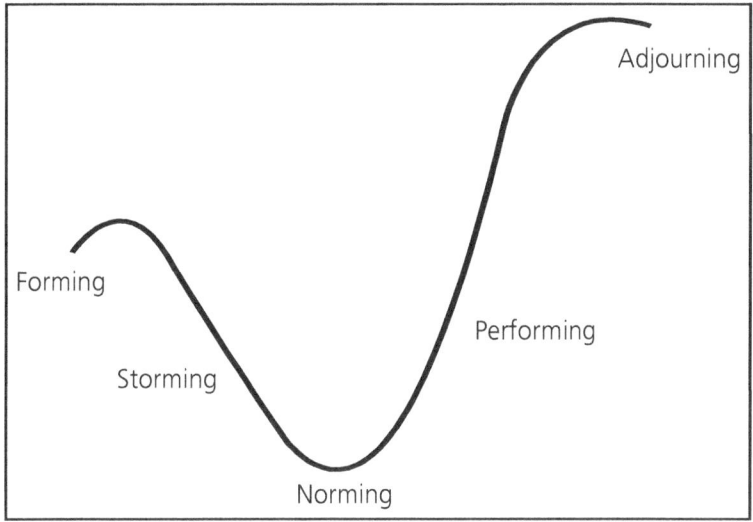

Figure 3-6. Stages in a Group's Growth

Storming. In this stage, the members begin to realize that achieving the group's objective is more difficult than they imagined. They realize that everyone has a different idea about how to accomplish the objective, but each person believes that his or her idea is the best. Group members often become testy, anxious, and argumentative. They rely solely on their personal and professional experience, and they resist collaboration with other group members.

There are so many pressures during the storming stage that group members are less productive. At this point, the focus should be on following the ground rules and sticking to their action plan in order to move on to the norming stage.

Norming. In this stage, the group begins to realize that not everybody thinks and acts the same way, which is probably good because if everyone thought and acted alike, life would be boring. More important, they realize that they can agree to disagree. The group members accept the individuality of other members and accept their roles in the group. Emotional conflicts are minimized as competitiveness turns into cooperation.

As members work out their differences, they find they have more time and energy to spend on achieving the group's objective. They start making significant progress. During the norming stage, the group requires less structure as they shift from conflict to cooperation.

Performing. In this stage, the group is effective, cohesive, and achieves positive results. Group members have settled into their relationships and have learned their roles. They have discovered and accepted each other's strengths and weaknesses. During the performing stage, group members should use creative problem-solving techniques and be cautious of groupthink. Groupthink is a term used to describe a team that functions so well together they begin to believe they are infallible. They readily accept statements and judgments coming from members within the group and lose sight of their critical thinking skills.[2]

Adjourning. You must let the group know they will disband at some point. The setup and changeover program doesn't end however, as another team composed of different members will focus on another product or process.

Knowing the typical stages that groups pass through should relieve much of the fear that members might have about whether their group will succeed. They should take some comfort in knowing that all groups experience these cycles of ups and downs.

---

2. Irving, Janis, *Victims of groupthink*, (Boston: Houghton Mifflin, 1972); Irving, Janis, *Groupthink: Psychological studies of policy decisions and fiascos*, 2nd ed. (Boston: Houghton Mifflin, 1982).

Challenging the Way You Do Things (20 to 30 minutes)

Paradigms are customs, traditions, or rules that people grow accustomed to in their lives, both at home and at work. People can become so comfortable with those customs, traditions, or rules that they simply accept things as they are, thereby missing important opportunities for change. The customs, traditions, and rules that bring comfort, logic, and order to our daily lives can actually inhibit us from solving problems and seeing new opportunities.

For example, some engineers at a Swiss watchmaker designed the quartz crystal.[3] However, that company's top managers decided not to make quartz crystal watches because they considered themselves craftspeople. Their paradigm was that a fine watch had to have lots of gears and hands that moved. As a result, the Swiss engineers displayed the quartz crystal in a common marketplace filled with industrialists. The quartz crystal was quickly picked up by Texas Instruments, which ran with it. Now the quartz crystal watch has the lion's share of the market, while the market share for Swiss watches is very small.

> ### SETUP REDUCTION DO'S AND DON'TS
>
> **Do**
>
> ✔ Answer the question "What's in it for me?" for everyone involved.
>
> ✔ Make your actions match your rhetoric to demonstrate your commitment to the group and the program.
>
> ✔ Encourage the group to challenge the status quo.
>
> **Don't**
>
> ⊗ Use a lot of hoopla when you explain the quick setup program and its benefits.
>
> ⊗ Tell the group members that your goal is to cut costs.

Another paradigm concerns the phrase "Made in Japan." In the 1960s, the phrase "Made in Japan" was equated with poor quality because at that time, Japanese manufacturers made cheap transistor

---

2. *Joel Barker's The New Business of Paradigms* DVD (St. Paul, MN: Star Thrower Distributions).

radios and small, junky toys. When the Japanese industrialists became aware of this negative paradigm, they did their homework and worked hard to improve their work processes and habits. As a result, the Japanese manufacturers not only improved the quality of their products but also changed the paradigm. Now when you talk about "Made in Japan," you are talking about high-quality products.

As these examples illustrate, companies must identify the paradigms that influence their business; otherwise, they might follow in the footsteps of the Swiss watchmaker. Similarly, the setup reduction group must identify the paradigms that influence the setup operation they are studying; otherwise, opportunities for improvement might be overlooked or dismissed as too unconventional. Thus, you need to ask the group members questions such the following:

- What paradigms exist in our setup practices?
- What can we do differently?
- What if we had the tools and materials right next to the machines?
- What if we had holding fixtures?
- What design changes would make the product easier to manufacture?
- What would happen if we could utilize different equipment?

Identifying and talking about paradigms helps set the stage for breaking away from conventional thinking—for challenging the status quo. For example, suppose that part of a setup operation involves turning a part over to the inspection lab and waiting for them to perform the first-article inspection. A group member might ask whether the inspection is necessary, to which the response is, "Yes, I think it's an ISO requirement." At that point, the group member can challenge the status quo by asking, "Does ISO really require us to do that? Couldn't the inspector come over to the machine? Or is that procedure something that we've just grown accustomed to doing?" Using an example to illustrate the paradigm gives group members a way to challenge conventional thinking.

POINTS TO REMEMBER

☛ Focus on what's in it for the employees

☛ Capture expectations and concerns using their terms and not yours

☛ Explain the dynamics of teamwork and problem solving

☛ Set ground rules to ensure success

CHAPTER 4

Establishing the Baseline

A vital step in the setup-reduction process is gathering information about the operation being studied so that the group can establish a snapshot, or *baseline*, of current practices. The group later uses this baseline to measure the success of its improvement efforts.

Be forewarned that the videotape will capture not only the good but also the bad and the ugly.

You can collect baseline information by having the setup-reduction group observe and document a typical setup process. However, for complicated setups, the best way to collect the baseline information is to videotape a setup operation because videotaping provides objective data that the group can review in detail.

How to Create a Video Document

Be forewarned, though, that the videotape will capture not only the good but also the bad and the ugly. The group will also need to document information about the setup operation. Before the group videotapes and documents the operation, the group leader needs to assign several roles to the group members.

The Cast of Characters

The group leader needs to make sure several roles are filled during the videotaping of the setup operation. A good place to start is to ask for a volunteer for each role. If there isn't a volunteer, you need to pick someone. The roles to fill are:

Setup operator(s). The setup operator(s) performs the setup operation in front of the video camera. This individual should be the person who normally works on the setup operation being studied. This person might be a dedicated setup operator or a machine operator who also performs the setup operation.

Videographer. The videographer videotapes the setup operation, including the teardown of the previous job, the setup of the new job, adjustments, inspections, and scavenger hunts for parts, tools, material, and people. It is important that the videographer be a setup or machine operator. Don't put the video camera into the hands of an engineer, manager, or supervisor. If you do, operators will then view the project as a time-and-motion study and may feel threatened.

Assistant videographer. Prior to the videotaping session, the assistant videographer assists with preparations, such as making sure there is ample lighting at the machine being filmed, a tripod for the video camera, extra video camera batteries that are available and fully charged, and a long extension cord in case all of the batteries run dry. During the videotaping session, the assistant videographer performs such tasks as changing the video camera's battery and taking over filming when the videographer needs a break. The assistant videographer also serves as a safety observer. For example, he or she makes sure that the videotaping doesn't interfere with the setup operator's safety and makes sure no one trips on the extension cord if one is used.

> **WARNING**
>
> Don't put the video camera into the hands of an engineer, manager, or supervisor. Operators will then view the project as a time-and-motion study.

Timekeeper. The timekeeper uses a stopwatch to track how long each step in the setup operation takes. The timekeeper also keeps track of the elapsed time, which is the time it takes to perform the entire setup operation. The timekeeper tells the times to the scribe, who records them.

Scribe. The scribe takes notes about the setup operation and records the times that the timekeeper provides. To help the scribe with this task, you can copy the Setup Baseline Worksheet (Figure 4-1) and give it to the scribe to complete. In the worksheet's Part Name and Description box, the scribe needs to write down the name and description of the job for which the machine is being setup. If the organization tracks die or mold numbers, the scribe should write that number in the Die box. In the Element Description column, the scribe needs to explain the steps that occur during the operation. He or she also needs to number those steps in the Element Number column. The amount of time each step takes is recorded in the

Setup Baseline Worksheet					
Date	**Part Name and Description**	**Die**			
Element Number	**Element Description**	**Duration (min:sec)**	**Elapsed Time**	**Internal Setup**	**External Setup**

Figure 4-1. Setup Baseline Worksheet

Duration column, whereas the elapsed time is written in the Elapsed Time column. During the videotaping, the scribe should leave the Internal Setup and External Setup columns blank. These columns will be filled in later.

Lights, Camera, Action

At this point, the group members have sat through two to three hours of discussion about what they are supposed to do and what to expect, so they are probably eager to put theory into practice and start videotaping and documenting the setup operation. When the group videotapes and documents the setup process, the setup operator and all other staff members should perform the setup operation the way they normally do. You don't want anyone to work faster. At the same time, you don't want anyone to drag out the job. Everyone should work just as they normally do.

Just before the group begins videotaping, the group leader should ask the group members if they have any questions about their roles. In addition, you should also spend some time addressing all necessary safety precautions: high pressure hoses, pinch points, hot surfaces, eye and hearing protection, etc. Emphasize the importance of staying out of the setup operator's way while he or she works. It is worth taking those extra minutes to underscore that you want the session to be a safe learning activity.

Typically, the videotaping session runs smoothly, with only minor glitches (e.g., a dead battery) that the group is usually prepared to handle. One question that might come up, though, is what to do if it's lunchtime and the setup operation is still being videotaped. People normally don't work through lunch. Because the group members have been told to perform the operation the way they normally do, they should take their lunches. Because nothing will be happening during that time, the videographer can turn off the video camera. To accurately account for this time, though, the scribe must add a step that reads "Lunchtime" and record the time spent at lunch so that the group gets a true reading of the operation's elapsed time.

Another question that might arise is what to do if a setup operation involves a long wait period. Suppose, for example, a group is videotaping an operation that involves heating a die on a heating plate for three hours before the die goes into the machine. Do the group members have to watch the videographer film the die sitting on a heating plate for three hours? Fortunately, they don't. Instead, the videographer can turn off the video camera with the timekeeper noting that time. The group members can then go back to the conference room and work on other tasks. However, the group must not assume that the die will take three hours to heat and come back three hours later. If the group does that, it won't know exactly when the die reached its operating temperature—the operating temperature might have been reached at two-and-a-half hours or just a minute before the group members returned. Rather than coming back three hours later, the group must periodically check on the die's progress. When the die is close to reaching the operating temperature, the videographer should again start filming the operation.

When the die reaches the operating temperature, the timekeeper should note the time. The timekeeper must then calculate how long the group waited for the die to heat up by comparing the time that the video camera was shut off against the time the die reached its operating temperature. After receiving this information from the timekeeper, the scribe needs to document the step "Waiting for the die to heat up" and record how long the step took in the Duration column as well as total time so far in the Elapsed Time column.

Not videotaping lunchtime breaks and long wait periods is the exception rather than the rule. The group needs to be selective in what it leaves out of the videotape. If there's a wait period that's going to be 20 or 30 minutes long, the group should keep the videotape rolling and the stopwatch running.

After the group finishes videotaping and documenting the setup operation, everybody should return to the meeting room. You should explain that they are going to watch and discuss the videotape and review the Setup Baseline Worksheet, but first let the group members take a break to freshen up and grab a soda or coffee. Many times, they have been working pretty hard. While the group members take their break, you should set up the VCR and get a flipchart ready so that you can take notes as the group watches the videotape. You should also make photocopies of the Setup Baseline Worksheet that the scribe filled out so that the entire group has a copy of it.

Figure 4-2 shows what a filled-in worksheet might look like. This worksheet was created by a setup reduction group in a company that manufactures pigment and chemical dispersions for thermoset plastics. The group videotaped and documented a process that blends different color plastics into custom colors, pelletizes the custom-colored plastic, dries the pellets, then drops the pellets into containers for shipping to manufacturers that extrude the pellets into products such as automobile dashboards. Figure 4-2 is only a portion of the entire worksheet, which included more than 75 steps.

What to Look for on the Videotape

When everybody is back in the conference room, you should remind the group members that the purpose of the videotape isn't to criticize any individual, and that it is just a tool to explore what's taking place during the setup operation. You should also explain that the group's job is to better understand the operation and look for possible improvements.

A group member should run the VCR. That way, you can make sure that the group members are discussing the videotape and you can write down the group's comments and ideas on the flipchart. Often there is a lot of background noise on the videotape, so the volume should be turned way down or off.

After a minute or so watch of watching the videotape, you should start asking questions that will not only elicit pertinent data about the setup operation but also get the group thinking about and openly discussing the operation. The types of questions that you should ask include:

- What was the total elapsed time for the setup operation?
- Was there anything out of the ordinary about this operation, or was it a typical setup?

Setup Baseline Worksheet					
Date 03/04/xx	Part Name and Description Part 27, Line 2	Die From: 32366 To: 40212			
Element Number	Element Description	Duration (min:sec)	Elapsed Time	Internal Setup	External Setup
1	Turn off extruder	0:15	0:15		
2	Unhook pelletizer	0:22	0:37		
3	Get purge	0:38	1:15		
4	Purge extruder	0:59	2:14		
5	Shut off screener	0:11	2:25		
6	Get forklift	0:10	2:35		
7	Move bin of processed product	1:00	3:35		
8	Get paperwork while bin is unloading	1:32	5:07		
9	Move dirty collection bin	0:23	5:30		
10	Move tote to weight scale	0:55	6:25		
11	Weigh and mark tote	0:39	7:04		
12	Remove tote and prepare tumbler	1:39	8:43		
13	Tumble tote	4:02	12:45		

Figure 4-2. Setup Baseline Worksheet (excerpt) for an Operation that Creates Custom-Colored Plastic Pellets

- How much scrap was created? How much setup material was used? Frequently during the trial run and adjustment period, before the setup operator gets the first good piece, a number of pieces have to be run off. This scrap is sometimes recorded in terms of the number of pieces; other times it is recorded in terms of the number of pounds of materials used.

- What's going on now? You should ask what is occurring at key steps in the setup operation. If needed, the setup operator who performed the operation can explain what he or she is doing in the videotape.

If any defensive posturing arises during the discussion, you need to end it immediately in a constructive way. For example, suppose that on the videotape, the setup operator leaves the machine and walks to a tool crib to get a wrench, and a group member comments, "Wouldn't it be easier if you just kept the wrench in your work area instead of walking to the crib?" Although the comment may not have been intended as an insult, the setup operator might take it that way and reply, "Now, wait a minute, it's not my idea of fun—there aren't enough tools to go around." If you sense that someone is being defensive, you need to chime in and say something like, "Remember that right now, we're only trying to capture the way the operation is currently performed

SETUP REDUCTION DO'S AND DON'TS

Do

✔ Spend a couple minutes talking about safety before videotaping to underscore that you want the session to be a safe learning activity.

✔ Have everyone work the way they normally do during the videotaping.

✔ Immediately end any defensive posturing that comes up during discussions of the videotape.

Don't

⊗ Be too eager to turn off the video camera during the setup operation. Except for lunchtime breaks and long wait periods, the video camera should keep rolling.

⊗ Watch every second of the videotape.

and identify areas for improvement we might want to explore. Later, we'll start to come up with some ideas on how we might do things differently. At that point, if there are reasons why we can't do them differently, you'll need to say so."

At some point, the videotape will capture a segment in which nothing is happening. When the group members watch their videotape for the first time and they reach a point where nothing is going on, leave the videotape running on normal play. Take a close look at the group members. After the first minute and a half, you will see the frustration in their body language and eyes. You can almost hear them thinking, "What in the world? What's going on? Why aren't we doing anything?" If you let them continue watching nothing for three or four minutes, people typically start to get upset, and, at that time, you should tell the VCR operator to fast-forward that part of the tape. While that's being done, start a discussion about what the group members just witnessed. Long dead times often make people realize why they need to reduce setup times, and this creates a sense of urgency for them to do so.

Another event that can change group members' outlook on setups is break time. It is common for employees to take their breaks. When group members capture breaks on videotape and watch that part of the videotape, they often become uncomfortable with the lack of activity and come away with an important realization: If break time happens to fall during a setup, the operators need to finish the changeover before they take their breaks.

When playing the videotape, it isn't necessary to watch every second. In fact, watching the entire videotape of a long setup operation is tiresome and might be counterproductive. Instead, the group should speed through those parts where all the action is behind the scenes. For example, suppose an operation involves putting several parts into a furnace that is vacuum enclosed. After the setup operator puts the parts in the furnace, the air is pumped out in order to create a vacuum. After the air is out, the furnace's temperature is slowly raised in 50-degree increments—a process called the heat-up cycle—so that any water in the furnace is vaporized at a safe temperature. (If water is vaporized at too high a temperature in a vacuum, an explosion will occur.)

Drawing out the air takes an hour, whereas the heat-up cycle takes several hours. While the air is being drawn out and the temperature is being slowly raised, all you will see in the videotape is the outside of the furnace. You don't need to watch the videotape during those segments because there's nothing to see—all the action is behind the scenes. However, the scribe does have to record how long it took for the air to be drawn out and the furnace to heat up. The scribe must then add these times to the elapsed time. Knowing how long these steps took and the total elapsed time can lead to thought-provoking questions in the next steps of the setup reduction program.

When the VCR operator is fast-forwarding through the footage in which the action is behind the scenes, you can start a discussion about what the operator was doing during that time and how that time can best be utilized to improve the setup process. For example, in the case of the vacuum-enclosed furnace, the operator might begin to prepare for the next job while the air is being drawn out and the temperature is slowly being raised.

The Next Steps

After the group finishes watching the videotape, there are usually a number of ideas written on the flipchart. To transform these ideas into implementable improvements, the group needs to separate internal activities from external activities, convert internal activities to external activities, streamline efforts, and standardize improvements.

POINTS TO REMEMBER

☞ Documenting existing setup and changeover practices establishes a baseline for measuring improvements

☞ Videotape is a good way to capture the good, the bad, and the ugly

☞ Emphasize safety when conducting setup and changeover

☞ Capture questions and ideas using a flipchart

☞ Keep the team focused on ideas for improvement and not judging/criticizing past behaviors and practices

CHAPTER 5

Separating Internal from External Operations

After the setup-reduction group has established a baseline for the operation it is studying, it needs to find ways to improve that operation; in other words, it needs to reduce the operation's total elapsed time. The best way to reduce that time is by first identifying which tasks are internal and which tasks are external in the Setup Baseline Worksheet, and, second, by making sure that only internal activities are performed while the machine is stopped. Before the group members can do that, they need to understand the difference between internal and external activities.

Difference Between Internal and External Activities

A setup operation consists of two types of activities: internal and external. Internal activities are those activities that the setup operators must perform while the machine is off, such as bolting a part onto a plate or adjusting a depth setting. External activities are those activities that the setup operators can perform while the machine is finishing the current job, such as getting the tools and materials ready for the next job. Sometimes this is called *out-of-press*.

When the chuck turns, the buck turns. The minute the chuck stops, the buck stops.

The idea behind internal versus external is to get as much of the next job's setup done while the machine is running the current job. You want to maximize a machine's running time because, as the saying goes, "When the chuck turns, the buck turns." The minute the chuck stops, the buck stops. The only time a chuck should stop turning is when the operator is exchanging parts on the machine or adjusting settings on the machine to get it to specification. Otherwise, the machine ought to be running.

However, that is not the case in many manufacturing facilities. Many tasks that could be done out-of-press—while the machine is running—aren't done until the machine is stopped. Take, for example, the before and after operations illustrated in Figure 5-1. Typically, the operation goes like this:

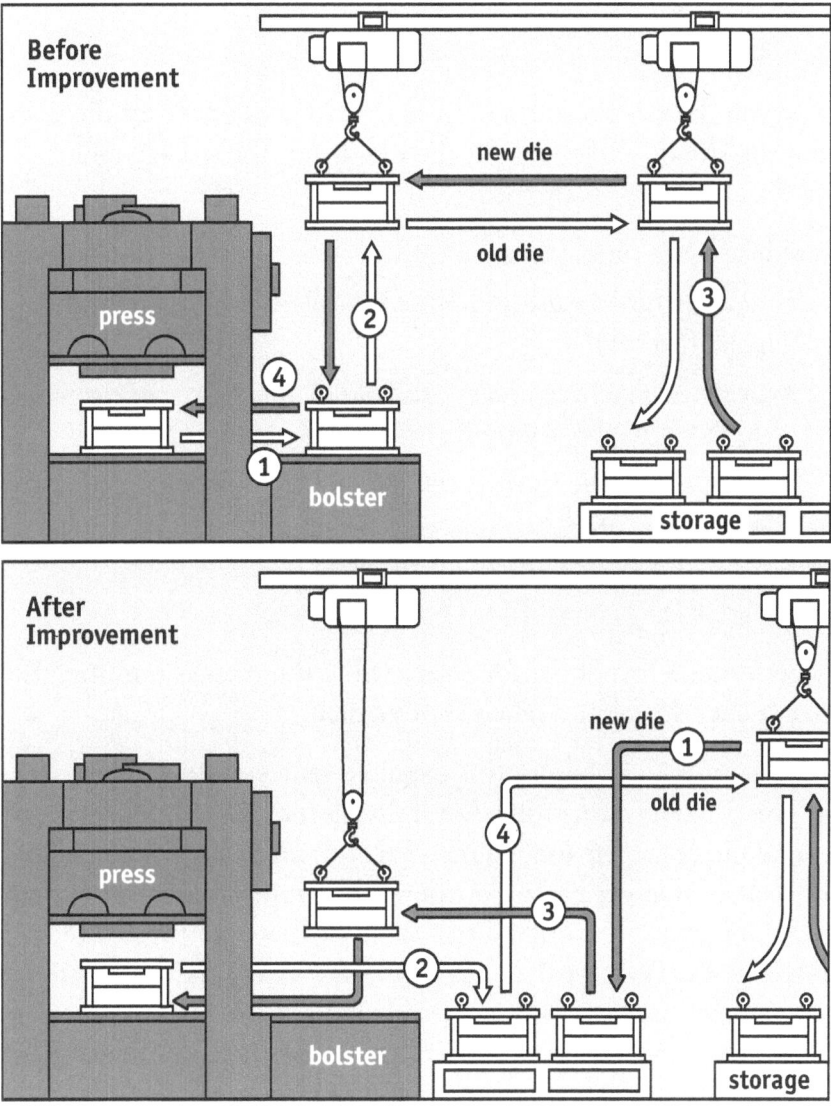

Reprinted from *Quick Changeover for Operators: The SMED System.* Productivity Press, 1996.

Figure 5-1. How to Reduce the Setup Time for an Operation: Before and After

- Stop the press, take the old die out of the press, and place it on the bolster (Figure 5-1. Step 1),

- Lift the die off the bolster with a crane if it's heavy and put the die on the storage table (Step 2),

- Lift the new die off the storage table and move it to the bolster (Step 3),

- Then put the new die in the press and start the press (Step 4).

However, there is no reason why you have to get the new die from the storage table and put the old die on the storage table while the machine is stopped.

To reduce setup time, you can:

- Go get the new die and place it as close as possible to the press (Step 1),
- Stop the machine and get the old die out of the press but don't worry about moving it to storage (Step 2),
- Put the new die in the press and get that machine running again (Step 3),
- Then put the old die away after the press is running (Step 4).

By performing two of the four steps while the press is running, the press is shut down for a much shorter time.

For one individual, Jason, this example struck close to home. In his press area at Armstrong Air, a division of Lennox, he and several die operators use a punch press to fabricate twelve different heat-exchanger parts. For them, the typical operation was to stop the punch press and go to a large rack, where they stored all the dies on wood pallets, as Figure 5-2 shows. After hunting for the correct die (dies weren't assigned a location on the rack), the operator used a forklift to pull the wood pallet and die off the rack and transport them to the punch press. The operator then manually transferred the die from the wood pallet to the press bed. Not surprisingly, this transportation system not only wasted valuable time but also was riddled with safety and ergonomic issues.

As part of a project to receive his OPMA Lean Enterprise Specialist certification, Jason and several die operators devised and implemented a new transportation system. (Jason was participating in the Lean Enterprise Specialist certification program offered by an Ohio-based group called Operations and Plant

Figure 5-2. Die Storage in the Old System

Manager's Association, or OPMA, which consists of individuals from noncompeting manufacturing companies who want to share best manufacturing practices. To become certified, participants must attend eight workshops, complete a lean manufacturing project in their companies, and present their projects at an OPMA meeting. As Figure 5-3 demonstrates, they now store the dies on a stationary table that has a roller conveyor on top. Each die has an assigned location for quick access. While the current job is running, the operator goes to the die storage area where he rolls the next job's die onto a transfer table. Like the stationary table, the transfer table has a roller conveyer on top. Unlike the stationary table, the transfer table is on wheels, so the operator simply pushes the transfer table to the press. When the current job is finished, the operator stops the press, rolls the current die off the press onto another transfer table, then rolls the upcoming job's die onto the press bed, as Figure 5-4 shows.

Figure 5-3. Moving the Die from the Stationary Table to the Transfer Table Using the New System

Figure 5-4. Moving the Die from the Transfer Table to the Press Bed Using the New System

Identifying Internal and External Activities in the Worksheet

After the setup-reduction group members review the videotaped setup operation, they need to analyze every step in the Setup Baseline Worksheet to determine whether it is internal or external. To make that determination, ask "Why do you do this?" for each activity on the worksheet.

Before you ask a group member a question such as "Why do you walk around so much?" you first need to explain that you are asking this thought-provoking question in order to find out whether there is a good reason for walking around or whether it is just a practice that everyone has been following for years. If you don't provide this explanation, you can expect people to get upset when asked such a question. They are likely to respond defensively, and say something like, "What are you saying—do you think I walk around because I want to?" By explaining why you are going to be asking "Why?" the group members will be a bit more at ease and will feel less threatened.

You must ask "Why?" because you must find out the reason behind each activity. Based on the response, the group can then decide whether the activity is internal or external. Figure 5-5, which shows part of the Setup Baseline Worksheet created by the setup-reduction group at the thermoset plastics company discussed in Chapter 4, offers a great example. As you saw in Figure 4-2, the group had filled in the steps in the setup operation and how long each step took. As Figure 5-5 shows, the group then identified the internal and external activities for each step in its setup operation. Based on this information, they were able to reduce the setup time from 83 minutes to 43 minutes, increase production capacity by 140,000 pounds per month, and save the company more than $100,000 each month.

Another determination the group needs to make is whether each activity in the Setup Baseline Worksheet adds value for the customer. Activities such as cutting metal and transforming metal into parts add value for the customer, whereas activities such as standing in line for an inspection and jury-rigging a tool don't add

Setup Baseline Worksheet					
Date 03/04/xx	**Part Name and Description** Part 27, Line 2	**Die** From: 32366 To: 40212			
Element Number	**Element Description**	**Duration (min:sec)**	**Elapsed Time**	**Internal Setup**	**External Setup**
1	Turn off extruder	0:15	0:15	X	
2	Unhook pelletizer	0:22	0:37	X	
3	Get purge	0:38	1:15		X
4	Purge extruder	0:59	2:14	X	
5	Shut off screener	0:11	2:25	X	
6	Get forklift	0:10	2:35		X
7	Move bin of processed product	1:00	3:35	X	
8*	Get paperwork while bin is unloading	1:32	5:07	X	X
9	Move dirty collection bin	0:23	5:30		X
10	Move tote to weight scale	0:55	6:25		X
11	Weigh and mark tote	0:39	7:04		X
12	Remove tote and prepare tumbler	1:39	8:43		X
13	Tumble tote	4:02	12:45		X

*This step is both an internal and external activity. While the bin is being unloaded (internal activity), the setup operator gets the paperwork for the next job (external activity).

Figure 5-5. Internal and External Activities: Setup Baseline Worksheet for an Operation that Creates Custom-Colored Plastic Pellets

value for the customer. The group should try to eliminate any non-value-adding activities.

How to Ensure that *Only* Internal Activities Occur When the Machine Is Off

After the setup-reduction group members identify the internal and external activities, they need to look for external activities that are being performed while the machine is stopped. If they find any, they need to determine how they can change the setup operation so that those activities are either eliminated or performed while the old job is running. For example, if the setup involves people who bring materials to the operator, such as a forklift driver who retrieves a die, the operator can tell the forklift driver ahead of time to bring the die. That way, the die can arrive while the machine is still running the current job.

If your company hasn't been working on improving setup time, a setup-reduction group can likely reduce an operation's setup time between 30 percent and 50 percent just by separating internal and external activities and making sure that only internal activities occur while the machine is stopped. For example, a setup reduction group in a small division of Carlisle, a Fortune 500 company, realized a 50 percent reduction in setup time for one mold press by doing just that.

Example 1: Locating Materials Efficiently

The customers of this small division in Canton, Ohio, are primarily automotive manufacturers, so the division churns out bumper end-caps, air manifolds, ventilation ducting, and other plastic-molded products at a high volume. Because of the large size of the moldings, blow molding is used instead of injection molding. Thus, the moldings usually have to undergo a secondary operation, in order to trim flash or drill holes.

The members of the setup-reduction group, which was lead by two OPMA-certified lean enterprise specialists, separated the press's internal and external activities. One time-wasting activity they immediately noticed was that the operators had to walk to different buildings to get what they needed for a job. Both the mold storage area and the fixtures that held the products during trim operations were in another building. In addition, the head tooling rack was at the other end of the plant. To eliminate this time-wasting activity, the group had the mold storage area, the trimming fixtures, and the head tooling rack moved to the press area.

Example 2: First-Article Inspection

Making an operator walk long distances to get necessary supplies can dramatically increase setup time. Another activity that can make setup time skyrocket is waiting for a first-article inspection. Consider, for example, the experience of one setup-reduction group at a company that welds and fabricates metal used in heavy equipment, such as off-road vehicles. The group was working on reducing the setup time for a welding operation. Typically, the setup time took about 30 minutes. After the welder finished a job, he would go get the new job order and the new fixtures and tooling, then weld the first piece. First-piece inspections were required, so the welder would notify the inspection department that he had a first piece to inspect, and then wait for an inspector to perform the inspection. However, first-piece inspections had no priority; instead, products that were going to be shipped that day were the top priority. Therefore, the welder typically waited for the first-piece inspection for about half an hour—and sometimes a lot longer.

After videotaping and documenting the process and identifying the internal and external activities, the setup-reduction group realized that some of the setup operation's external activities (e.g., first-piece inspection) were occurring while the "machine" was stopped—in other words, while the welder wasn't welding. So, the group came up with a better way to perform the setup operation. When the welder was almost finished with his current welding job, he would stop and get the new job order and the new fixtures and tooling, and then weld the first piece of the new job. He would then notify the inspector that the piece was ready for first-article inspection by drawing a curtain. While the inspector would perform first-piece inspection, the welder would go back and finish the previous job. Assuming that the first piece was approved, the welder could start the new job right after he finished the old one.

To carry out this plan, the setup-reduction group made a couple of changes to the welder's work area. First, the group installed a second welding table. Second, the group set up a curtain between the two welding tables so that the inspector could perform the first-piece inspection for the new job while the welder finished welding the pieces in the old job. (The curtain is required to protect the inspector's eyes.)

> ## *SETUP REDUCTION DO'S AND DON'TS*
>
> **Do**
>
> ✔ Make sure the group understands the difference between internal and external activities.
>
> ✔ Get as much of the next job's setup done as possible while the current job is running.
>
> **Don't**
>
> ⊗ Forget to explain why you'll be asking "Why do you do this?"
>
> ⊗ Be afraid to let management take care of a problem that the group identifies if it can do so quickly and easily.

More important than the work area changes, though, was the change that the group made to the inspectors' priorities. First-piece inspections now received top priority.

The new setup operation has virtually eliminated welding setup times, and although the setup reduction group had implemented only a few simple changes, the ROI is no small chunk of change. The company estimates that it is saving $187,500 per year.

Example 3: Establishing Standardized Procedures

An automotive components manufacturer is experiencing similar savings because of the efforts of a setup-reduction group working on a pinion forging press. The changes implemented by this group are saving the company more than $80,000 annually and have helped the company avoid spending $950,000 on capital equipment.

The initial videotape of the press's setup operation revealed that the setup took around two hours. When the setup-reduction group started identifying the internal and external activities, they discovered that there was no official setup procedure in place. Each die setter created his own procedure to set up the 2,000-ton mechanical press, which had two die stacks for each part. The group also uncovered other problems:

- The dies were prepared and stored in a separate die room. Because the die setter wasn't called in to set up the new job until after the old job completed, the die setter had to make a trip to the die room while the press was stopped.

- To run the press, the dies had to be heated. Although this company's die room would preheat the die on heating plates to save setup time, it did so only when it was convenient. If the die wasn't put on a heating plate, the die had to heat up while the press was stopped.

The setup-reduction group realized it had to address all these problems, and created a standardized procedure that all the die setters and other employees (e.g., press operators) must follow. In this new setup operation, the die setter is notified about a new job 15 to 20 minutes before the current job is finished, and before that job is completed, the dies are retrieved from the die room and placed on a heating plate right next to the press. Other needed tools and materials are also staged at the press. These changes led to a significant reduction in setup time—from two hours to an average of ten minutes (with a best time of six minutes)—without significant capital investment.

As these three examples show, the dedication and efforts of the setup-reduction group pay off both literally and figuratively. Sometimes, management can immediately take action on a problem that the group identifies. For example, when one setup-reduction group at Flo-Tork, Inc., watched its videotape, the group saw many operators walking around looking for vises. When one of these operators was asked why he was walking around, he replied that he wanted to put a part in a vise on the machine. However, his production area had only two vises for six machines, so he was hunting for one of the two vises. When management found out about the need for more vises, it quickly determined that the cost associated with lost machine time and the benefit associated with having less frustrated employees and higher-quality parts easily justified the $4,000 price tag for four new vises, which the company purchased. This immediate follow through not only helped improve setup times but also demonstrated management's commitment to the setup reduction program.

POINTS TO REMEMBER

☛ Separating internal activities from those that are external can reduce setup and changeover time by as much as 50 percent.

☛ Planning for the next job goes a long way toward reducing time for setup and changeover.

☛ Setup operators, tool setters, and machine operators have ideas and suggestions for reducing setup and changeover. Listen to them and follow through immediately to show that you are serious.

CHAPTER 6

Converting Internal to External Operations

Whenever possible, the setup reduction group should convert internal activities to external activities. When you separate internal activities from external activities, you are not changing the activities in the process. You are just changing *when* you do some activities so that more activities are done while the machine is running. When you convert internal to external, however, you are actually changing an activity in the process.

For example, after a race car gets new tires, the driver wiggles the car back and forth to warm up those tires. In an effort to convert an internal activity to an external one, the pit crew could use an oven to preheat the tires so that after the new tires are put on, the driver doesn't have to wiggle the car and can get to racing speed faster. Now, NASCAR might not approve of this idea, but it illustrates how the pit crew can convert an internal activity to an external one.

Ways to Convert

Unlike the NASCAR pit crew, your setup-reduction group might actually be able to apply this idea. Let's say that in order for a press to operate properly, the die has to be at 450° Fahrenheit. Most companies would take a room temperature die and insert it on the press, turn on the machine, and let the die warm up to 450°, which might take an hour. (The automotive components manufacturer discussed in Chapter 5 is one of the few manufacturers that preheats dies.) By putting the die on a heating plate and heating the die to 450° before inserting it in the press, you can reduce the setup time by that hour. In addition, preheating dies often reduces the number of trial runs and adjustments as well as the amount of setup scrap. Preheating dies is one way you can convert an internal activity to an external one.

Another way to convert internal to external is to standardize the sizes of the dies. When dies have different heights and widths, the operator must perform numerous adjustments on the machine, which is called shut-height adjustment. To standardize die sizes, the setup operators can add shims to the die every time they use it. Eventually, the operators will create a standard so that the dies are all the same height and width.

Converting internal activities to external activities typically takes more effort than separating internal and external activities and making sure that only internal activities occur while the machine is stopped. However, converting internal to external can lead to significant reductions in setup times and related benefits, such as increased production capacity. For example, an electrical wire producer was able to reduce the setup time for extruding silicon-coated wire by 33 percent, which increased production capacity by six million feet per month. A high-tech ceramics manufacturer reduced the setup time for one line so much that production on that line increased by 12 percent. Even more remarkable, a manufacturer of water treatment equipment reduced the setup time for one press by 75 percent. How did they do that? Let's take a look.

Example 1: Getting Down to the Wire

The setup-reduction group at the electrical wire company decided to reduce the setup time for a line that created silicon-coated wire. In that line operation, the wire went through extrusion, after which it received a silicon coating. The coated wire then went through a printer, which printed a part number and an approval number every foot or so. (The approval number was proof that the wire met certain standards, such as the Aircraft Wire Gauge, or AWG, standard.) The part number changed for each run, so during the setup operation for the new job, the operator took the print wheel out of the printer, put in a new print wheel, then adjusted the print wheel until it printed cleanly (i.e., not too much ink—so it wouldn't smear, and not too little—so it couldn't be read). While the operator changed and adjusted the print wheel, the wire ran through the printer at a very high speed. Not surprisingly, this setup process took a lot of time and produced a lot of scrap.

After analyzing the situation and a little experimentation, the setup-reduction group discovered that it would be faster to change the entire printer stand rather than just the print wheel. That way, the operator could adjust the print wheel off-line. In addition, converting this internal activity to an external one would not cost much because the company already had an extra printer stand.

The setup-reduction group implemented its idea. Now when the current job is running, the operator gets the next job's paperwork (which includes the job number) and print wheel. The operator puts the print wheel in the off-line printer stand and adjusts the print wheel using a sample piece of wire. When the current job is finished, the operator simply rolls the new printer stand in place and the printer is ready to go. This conversion from internal to external reduced the setup time by about 20 minutes (a 33 percent reduction) and reduced the process scrap from 7 percent to 4 percent.

Example 2: A Hot Situation

The setup-reduction group at the high-tech ceramics manufacturer was working on ways to reduce the setup time for an operation that produced a high-tech ceramic used in applications such as heat sinks for semiconductors. In this operation, the operators place boron nitride inside a furnace, draw a vacuum, then have the furnace go through a slow heat-up cycle so that any water in the furnace is vaporized at a safe temperature. When the furnace reaches its target temperature, they introduce gases that deposit gas vapors on the boron nitride.

The operators had to wait a long time for the furnace to heat up to the extreme temperature needed (around 2000° Celsius). The setup-reduction group concluded that it could reduce the time that the heat-up cycle took by reducing the amount of moisture in the furnace. The group discovered that it could keep the moisture to a minimum by filling the furnace with nitrogen at the end of each run and by keeping the furnace door closed as much as possible. The operators had been leaving the furnace door open when they loaded or unloaded the furnace. Now when they're building the load stack, they keep the furnace door closed. In addition, they close the door as soon as they unload the furnace.

Example 3: The Creative Connection

Kinetico—a manufacturer of water treatment equipment for residential, commercial, industrial, and municipal applications—found out firsthand what a little ingenuity can lead to. In Kinetico's Consumer Products Group, the setup operations for presses in the molding area were taking a long time. One particular press, known as Press #8, had an average setup time of about one hour, but on occasion, setups took as long as four hours. So, a setup-reduction group decided this press was a good candidate for an improvement effort.

After providing training for that press's setup operators, the setup reduction group members started measuring and tracking the time every setup operation took. Using problem-solving tools, such as Pareto analysis, they methodically identified the factors contributing to the long setup times, then eliminated or reduced the effect of those factors. For example, one of the group's most notable improvements dealt with the water lines flowing between the press and mold. As illustrated in Figure 6-1, the press originally had a manifold with eight water connections, which meant that the setup operators had to uncouple, then couple eight connections each time they changed a mold.

Uncoupling then coupling the connections had to be done while the press was stopped, so there was no way the setup-reduction group could make this an external activity. However, the group members were unwilling to shrug their shoulders and say, "Oh well, let's move on." Instead, they came up with an innovative plan of action: design and build a new manifold. As Figure 6-2 shows, the new manifold has only two connections—one for incoming water and one for outgoing water—which significantly reduces the amount of time the setup operator spends on uncoupling then coupling connections. Although technically the group didn't convert the internal activity to an external one (they converted the lengthy internal activity to a much shorter internal activity), they did change the activity by designing and building a new manifold.

Figure 6-1. The Old Manifold

Figure 6-2. The New Manifold

From the onset of its efforts, the setup-reduction group had charted the setup times for Press #8. As Figure 6-3 demonstrates, the press's average setup time has steadily decreased.

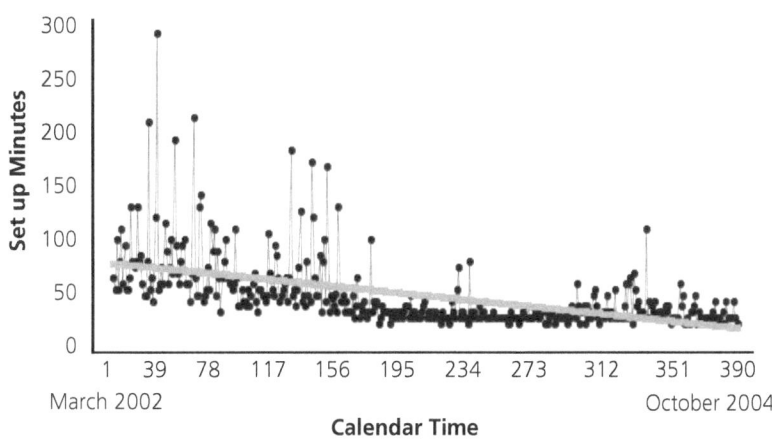

SMED Program—Press #8 Tank to Tank (3/2002–10/2004)
$Y_t = 58.2931 - 0.147585*t$

Figure 6-3. Average Setup Times for Press #8

Happy with the results for Press #8, more setup-reduction groups were formed to reduce the setup times for other presses in the molding area. Figure 6-4 is a snapshot of the average setup times

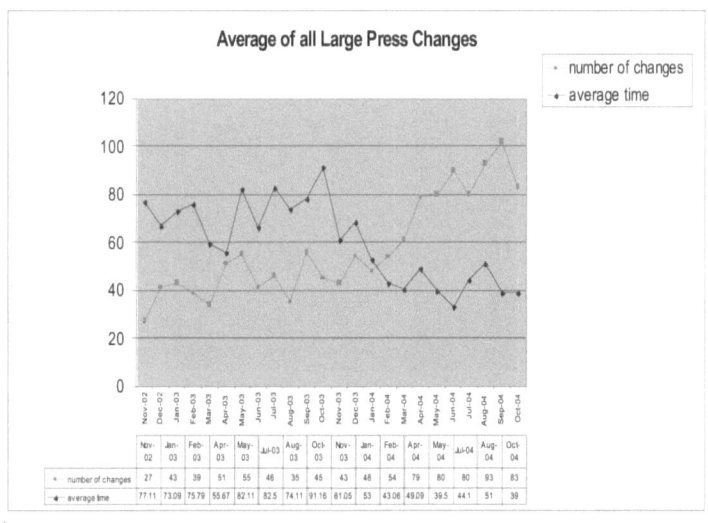

	Nov-02	Jan-03	Feb-03	Apr-03	May-03	Jul-03	Aug-03	Oct-03	Nov-03	Jan-04	Feb-04	Apr-04	May-04	Jul-04	Aug-04	Oct-04
number of changes	27	43	39	51	55	46	35	45	43	48	54	79	80	80	93	83
average time	77.11	73.09	75.79	55.67	82.11	82.5	74.11	91.16	61.05	53	43.06	49.09	39.5	44.1	51	38

Figure 6-4. Average Setup Times and Average Number of Setups for Large Press Changes on All Presses

for large press changes on all presses (not just Press #8). This graph also tracks the number of setups performed on these presses. As you can see, the average setup time is decreasing while the average number of setups performed is increasing. This graph illustrates an important reality: Reducing setup time naturally leads to the ability to perform more setup operations.

POINTS TO REMEMBER

- ☞ Converting activities from internal to external requires a little bit of creativity.

- ☞ It is helpful to provide examples, like preheating dies, exchanging printer stands, using quick connect/ disconnect fittings, to explain this concept.

- ☞ Measure how long it takes to complete various activities for setup and changeover.

- ☞ Prioritize the list and focus on reducing the longest time period first and then proceed to the next longest.

- ☞ Post the results so everyone can see progress.

CHAPTER 7

Streamlining Setup Efforts

At this point, the setup-reduction group has addressed the major causes of slow setup times. Now, it is time to examine and streamline the remaining internal and external activities. However, after working on the same project for some time, people can begin to run dry of ideas. To help the group, try using the Setup Analysis Worksheet (Figure 7.1). The questions in this worksheet can help generate discussions that can lead to further improvements.

How to Use the Setup Analysis Worksheet

The Setup Analysis Worksheet is simple to use. For each row, ask the group members to first answer the question in the left column, and then answer the follow-up question in the right column.

Take, for example, question No. 7. Suppose that the setup operator has to perform many tasks at both the front and back ends of the machine. Because there are many tasks, the group should brainstorm ways in which they might divide the setup operation between two setup operators. Running parallel operations is safe as long as the setup operators signal each other when they have completed their steps.

Some managers are reluctant to use more than one setup operator because it costs too much in labor time. However, these managers are missing an important point: You always use clock time, not labor time, to measure setup time. Clock time is more important than labor time because you can never recapture lost machine time. For example, if you have a cycle time of one piece per half hour and you shut down the machine for an hour, you just lost two parts. Because the company isn't making money when the machine is down, employees need to work together to get that machine running again, just like pit crew members work together to get their race car back out on the track.

How to Find Creative Solutions

While the questions raised in Figure 7-1 are certainly helpful in coming up with ways to streamline a setup operation, the list is by no means exhaustive. The group can probably come up with

Setup Analysis Worksheet	
1. What tools and tasks are needed to complete the setup?	Which of them should be included on a checklist?
2. On which parts of the equipment could you perform a function check during external setup?	How would you make sure the parts are in working order?
3. What parts and tools need to be stored and transported during external setup?	How could the storage and transport of parts and tools be streamlined?
4. What operating conditions need to exist to complete the setup?	Which of them could be prepared in advance?
5. Which parts of the equipment have to be changed or adjusted during setup?	Which of them would benefit from function standardization?

Figure 7-1. Setup Analysis Worksheet

(continued on next page)

Setup Analysis Worksheet	
6. Which parts or work pieces have to be centered or positioned after they are placed in the machine?	Which of them might benefit from the use of intermediary jigs?
7. What setup tasks require the operator to perform operations at both the front and back of the machine, or cause the operator to walk back and forth a lot?	How might you divide the setup operation into parallel operations?
8. What tasks require attaching dies, tools, or parts directly to the machine using bolts?	How might you replace bolts in the fastening procedures with functional clamps (e.g., one-turn, one-motion, or interlocking clamps)?
9. What trial runs or adjustments need to be made before the new operation can begin?	Could any of the trial runs or adjustments be eliminated by using fixed numerical settings, visible centerlines and reference plans?
10. Do any large dies or other heavy tools or parts need to be moved during the setup procedure?	How might moving these items be mechanized?

Figure 7-1. (*Continued*)

many ways to streamline an operation if they think creatively. For example, one way to streamline an operation is to analyze part design to see what changes could be made to simplify the part (e.g., more tolerance between pieces, easier tool path). Another way to streamline an operation is to standardize the bolts and clamps on the machines so that the operators don't need a lot of wrenches or other tools. Over time, it's not unusual for someone to lose a bolt when he or she is exchanging a plate. If extra bolts aren't available in the work area, the person will probably go on a scavenger hunt for a bolt. Although he or she will likely find a bolt that works, chances are that it will have a different sized head or a different length of thread. So, after awhile, operators will find themselves having to use two wrenches instead of one to do the setup—all because a few bolts were lost or damaged.

The same situation can occur with clamps. You'll often find many different types of clamping devices and methods used on the production floor. By streamlining this part of the operation, the group can not only standardize the bolts and clamps but also select the types of bolts and clamps that will work best for the operators. For example, the group might find that a quick release type of clamp can make the operators' jobs easier. Similarly, using a shorter bolt might mean that the operators need to turn the bolts fewer times. Because only the last three threads on a bolt hold an item in position, using shorter bolts isn't compromising the operators' safety or part quality.

SETUP REDUCTION DO'S AND DON'TS

Do

- ✔ Standardize the bolts and the clamps that are used.
- ✔ Place tools on shadow board or tool cart.
- ✔ Install quick connect/ disconnect fittings.

Don't

- ⊗ Keep a spare parts drawer full of odds and ends.
- ⊗ Store tools in separate areas.
- ⊗ Leave to chance the availability of a forklift or crane.

Save Time, Standardize Hardware

Standardizing the bolts and clamps might sound trivial, but it's not. Consider the following experiences of a heating and air conditioning products manufacturer and a flat wire manufacturer.

As you read in Chapter 5, an OPMA-Certified Lean Enterprise Specialist and several die operators at Armstrong Air applied the principles of setup time reduction to a punch press used to fabricate 12 different heat-exchanger parts. After making several major changes in the way in which dies were transferred from the storage area to the press bed, this setup-reduction group sought to streamline the use and storage of the clamps used to hold those dies.

At the time, the operators stored all the die clamps in one bin. As Figure 7-2 illustrates, this bin also contained washers used to shim up die clamps, old bolts, and miscellaneous tools.

Figure 7-2. Bin Originally Used to Store Die Clamps

To streamline this part of the operation, the setup-reduction team had special die clamps fabricated for each die, based on the die's thickness. These new die clamps eliminated the use of washers to shim up dies. To make it easy to find the correct clamp for a particular die, each die clamp has a designated storage bin, as Figure 7-3 shows. Each die clamp storage bin is labeled with the die part number. The new die clamps and storage system has not only reduced setup times but also operator frustration.

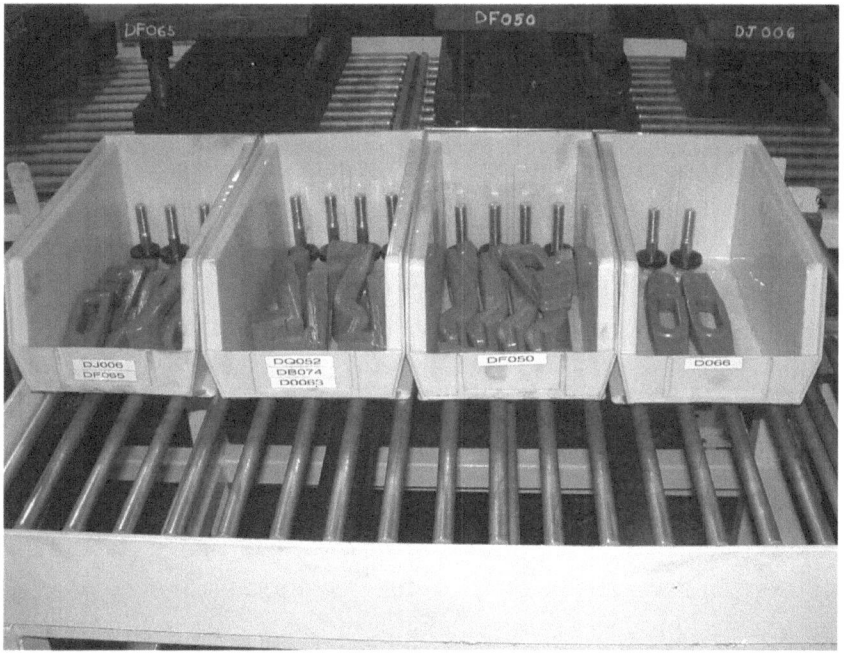

Figure 7-3. New Bin System Used to Store New Die Clamps

Armstrong Air isn't the only company to figure out that standardizing hardware is a worthwhile effort. One flat wire manufacturer that uses a multistage rolling unit to bend and shape wire has also discovered the benefits.

In the multistage rolling operation, a round thick steel wire goes through a milling machine and through as many as 12 different roller sets to change its shape and size. The wire comes out at the other end as a thin strip of steel that is then rolled up into a coil. The company's customers fabricate products from the steel strip. For example, a customer might cut it into short pieces and drill a hole in each piece to make brackets for mounting screws used to attach upholstery in automobiles.

When setting up this machine for a new job, the operators had to use a wrench to unscrew between four and eight bolts on a roller set, take the old roller off, put the new roller on, then use the wrench to tighten the bolts. With 12 roller sets, this process was time-consuming. The fact that the setup operators had to use multiple wrenches because the bolt sizes varied added time to the process and frustration for the operators.

When the setup-reduction group started to look for ways to streamline this setup operation, they discovered that it wasn't the bolts' tension holding the rollers in place. They realized that they could use bolts that could be finger tightened, and, therefore, they replaced all the bolts with standardized bolts that work like a thumbscrew. Because the setup operators can tighten and loosen these bolts by hand, they no longer need to use wrenches. The group also supplied neatly organized carts full of extra bolts, clamps, and other tools so the operators don't have to go on scavenger hunts if they drop a bolt or misplace a clamp.

Before the group's efforts, the setup operation took around 210 minutes and produced about 70 pounds of scrap. Since all of the improvements, including streamlining the operation by changing and standardizing the bolts, the setup operation takes only 90 minutes and produces only 65 pounds of scrap. As a result, the company is realizing an annual savings of $133,000.

Other streamlining techniques include marking dies and tools with color codes and location numbers to improve the storage. You can also eliminate adjustments in machines by using fixed numerical settings, visible centerlines and reference points, and mechanical stops.

How to Make "Dwell Time" Productive Time

After the group has done everything it can to reduce setup time, there still might be "dwell time"—unavoidable time that the operator must spend waiting. Instead of standing around, the operator can put this time to good use. For short dwell times, the operator might clean his area. For longer dwell times, the operator might help another department by deburring parts or even serving as an inspector in other production areas.

POINTS TO REMEMBER

☛ If you've ever watched a machine sit idle while someone searches for a misplaced bolt or wrench, then you know the importance of being organized.

☛ Use the same bolt size whenever possible to reduce the number of tools or wrenches needed. It makes the job easier for operators and simplifies the job too.

☛ Use shadow boards or tool carts to keep setup and changeover tools organized.

☛ Label bins and drawers so clamps and bolts are easy to find.

☛ Use quick release clamps and one-touch screws to save time loosening and tightening threaded bolts.

CHAPTER 8

Standardizing Improvements

With all the improvements in place, it's time to make certain that the improvements stick. By comparing the new setup time with the baseline, the group will know whether its improvement efforts were successful.

On occasion, setup reduction groups stub their toes and don't achieve the expected results. However, these groups don't stop. Instead, they look at what went wrong, figure out what to do differently, and try again.

If the setup reduction efforts were successful, the group needs to:

- Document the new procedures by creating a setup checklist.
- Provide training on the new procedures.
- Create a punch list.

Always remember that major improvements in one area are most likely transferable elsewhere. The setup-reduction group members should actively search for other setup operations that might benefit from the improvements they made in the original setup operation.

Creating a Setup Checklist

The setup-reduction group needs to document the new procedures for the setup operation by creating a setup checklist that all setup operators can use. There is no standard form or format that the group must use when creating the checklist. The only requirement is that the checklist be easy to understand and use.

Figure 8-1 shows a very simple checklist created by a setup-reduction group at a rubber extrusion company. The group implemented several improvements to reduce the setup time for its extrusion machines. The improvements included bringing containers of raw rubber strips closer to the machines for easy access, providing the setup operators with tool carts that contain the tools and dies they need, and purchasing an additional die so that the setup operator didn't have to clean a particular die from the previous job before he set it up in the new job. After the improvements resulted in a 60 percent reduction in setup time (from 45 minutes to 18

Rubber Extrusion Setup Checklists

External Checklist
Get process sheet envelope
Get material
Get die and screen
Get labels and staple gun
Get skids and packaging
Get the 10-to-1 drawing
Get the 10-to-1 comparator and make sure it works
Get the template and tools
Get the opening die and the air die

Internal Checklist (Operator)	**Internal Checklist (Helper)**
Set zones	Change/review cutter
Die placement	Change length on cutter
Feed stock	Set puller speed
Heat die	Install opening die and air die
Set conveyor speed	Set weight scale
	Set 12-inch cut fixture

Figure 8-1. Rubber Extrusion Setup Checklists

minutes), the group created the external and internal setup checklists shown in Figure 8-1. The external setup checklist outlines the tasks to be completed before the machine is shut down. The two internal setup checklists outline the tasks to be completed while the machine is shut down. The operator and the helper both work on their respective internal checklists simultaneously.

Figure 8-2 shows a more sophisticated setup checklist created by a setup reduction group at a tooling manufacturer. The group implemented several improvements to reduce the setup time for a machine that uses a series of diamond cutting wheels to cut and shape solid carbide blanks into an end mill or other tool that job shops use for machining. After the group's improvements reduced the setup time by 30 percent (from 99 minutes to 70 minutes), the group created the checklist in Figure 8-2.

Carbide Cutting Machine Setup Checklist

When machines are down, we are losing money. It's important to setup and changeover from the last complete tool to the first complete tool in a safe and efficient manner. Use the following checklist as a guide for setup and changeover.

Complete the following steps while the last job is running:

Gather program file and offset sheet ☐
Obtain:
    Wheel stack ☐
    Nose cone ☐
    Steady rest ☐
    Tools to do the changeover ☐
    Scrap piece for end work ☐
Fill out previous job offset sheet ☐
Edit data file—input flute wheel diameter ☐
Create program ☐
Create subprograms ☐
Transfer paperwork and tools from shelf to machine ☐

Once the last job is complete, perform the following steps:

Download all programs ☐
Remove previous job parts (nose cone, steady rest, etc.) ☐
Mount:
    Fluting wheel ☐
    Nose cone ☐
    Steady rest ☐
Set front stop locator according to the program ☐
Change offsets ☐
Dry run and position U axis ☐
V-cal if necessary (ball tool) ☐
Run end work on scrap piece ☐
Check:
    Web dimensions ☐
    End stock removal ☐
    Primary axial land width ☐
    Gash depths ☐
    Primary end wheel (coming to center) ☐
Make adjustments ☐

Figure 8-2. Carbide Cutting Machine Setup Checklist

Providing Training

After the group members document the new procedures in the setup checklist, they need to provide training for everyone affected by those procedures. The training must emphasize the need to follow the new setup operation. You need to make sure that people follow the new procedures so that the improvements don't disintegrate. However, that doesn't mean there isn't room for improvement. In the training session, the group must let all the participants know that their suggestions are welcome. Upon receiving a suggestion, the group should reconvene to review it.

Creating a Punch List

A punch list is a term that you often hear in the construction industry. For example, when a construction company finishes building a new house, the prospective homeowners walk through the house looking for anything that needs to be done before they sign a release form that tells the finance company to make the final payment to the construction company. Quite often, there are minor items that the construction company still needs to do, such as replace a cracked backplate on a light switch or touch up the paint on a wall. These things don't stop the homeowners from accepting the house, signing the release form, and occupying the house, but the homeowners still want the construction company to address those items, so they write them down in a punch list.

Similarly, the leader of the setup reduction group needs to create a punch list. The new setup practices that the group has standardized might not be all the improvements that the group has come up with. The group might not have been able to implement some of the improvements immediately, perhaps because new materials had to be ordered or an employee from another department needed to be involved. You don't want these unimplemented improvement ideas to be forgotten, so you need to put the unimplemented improvement ideas in a punch list. Someone must be responsible for each idea in the punch list.

For example, Figure 8-3 is a punch list that the setup reduction group at the rubber extrusion company initially created for its extrusion operation. Note that it includes not only the items to be completed and who is responsible for each item, but also a target date for each item's completion.

Sample Punch List		
Redesign die holder	Jason	By 03/05/xx
Repair puller/FPM indicator	Bill	By 03/08/xx

Tape machine—
    positioned blocks (two sets)
        Identify plates
	Dianne	By 03/08/xx

Accu. punch
    Relocate punch dies/tool tray
    Set screws
    Simplify adjustment for eyes
    and rollers
	Bill
	Dianne
	
	Bill/Dianne/Jason
		By 03/08/xx
		By 03/08/xx
		
		By 03/08/xx

Comparator sun shield	Jason	By 03/15/xx
Repair or replace lights	Bill	By 03/15/xx

Figure 8-3. Sample Punch List

It's been our experience that people often create a punch list and keep track of the items on it, but don't post the list. As a result, when the operators come to work each day, they don't see any progress. The group's punch list needs to be posted and visible to everyone. If you create the punch list on a flipchart and stick it on a prominent wall in the work area, all the operators can see that you made the punch list and they can see the progress as you cross off items on that list. This is one of the most powerful ways to communicate to people that you're really committed to making the improvements necessary for setup reduction. It helps answer the one question that is always on the group members' minds: Is management really going to follow through on all the improvements we recommended?

Another benefit of posting the punch list in a prominent location is that people are a lot less likely to try to back out of their responsibilities. When 13 out of 15 items are crossed off on the punch list and that punch list is visible by all, it puts a little peer pressure on those people whose names are next to the 2 items that aren't crossed off. At the same time, the pressure isn't personal or blaming in nature.

SETUP REDUCTION DO'S AND DON'TS

Do

✔ Create a punch list of items to be completed.

✔ Post the punch list in a conspicuous place.

✔ Cross off items when they're completed.

Don't

⊗ Keep the punch list in a folder or notebook.

⊗ Ignore uncompleted items on the punch list.

⊗ Wait for all the items to be completed before communicating the results.

POINTS TO REMEMBER

☛ Use flipcharts or white boards to record and track action items left unfinished from setup reduction efforts.

☛ Document improved setup and changeover practices using checklists and procedures.

☛ Provide training so everyone follows the new setup and changeover practices.

☛ Videotape the new setup procedures and use the tape for training.

☛ Publicize new practices so employees in different departments or plants don't reinvent the wheel.

CHAPTER 9

Celebrate Successes

Management needs to recognize and celebrate the group's success very publicly. Perhaps more important, management needs to brag about the group's success to other departments. Manufacturing managers might not be marketers, but if there is one thing they need to market, it's this program. You want other employees to know the benefits of the setup-reduction program because you want them to start using those ideas.

You Get What You Reinforce

Aubrey C. Daniels, a noted psychologist and author of *Bringing Out the Best in People*, states that "you get what you reinforce."[1] In other words, the behavior your employees display at work is the behavior your organization reinforces. To change employees' attitudes and habits, you must start by changing the consequences of their behavior. To put it simply, people behave in ways that give them positive reinforcement. Positive reinforcement causes people to repeat a behavior because of a pleasant result. If you compliment an operator for catching and fixing a problem rather than passing it downstream, he or she will continue to be vigilant. Negative reinforcement causes people to repeat a behavior to escape an unpleasant result. If you chew out an operator for missing a problem, he or she will likely go overboard finding fault with everything and placing blame on everyone else.

There's a big difference between positive and negative reinforcement. Positive reinforcement makes people want to do better and invest more discretionary effort. Negative reinforcement makes people do only what is necessary to escape punishment. This difference leads to Daniels' Antecedents, Behavior, and Consequences (ABCs) of changing behavior.

The ABCs of Changing Behavior

Antecedents set the stage for behavior to occur the first time. Most organizations invest heavily in antecedents. Training classes,

---

1. Aubrey C. Daniels, *Bringing Out the Best in People* (New York: McGraw-Hill, 1999).

policies, procedures, memos, banners, and slogans are all examples of antecedents. For example, many organizations have posters saying: "Safety First" or "Quality Begins with Me." However, antecedents alone aren't enough. The behavior they tout must be followed by consequences. Unfortunately, organizations provide little positive reinforcement for either safety or quality. Instead, they rely

> **SETUP REDUCTION DO'S AND DON'TS**
>
> **Do**
>
> ✔ Brag about the group's success to other departments.
>
> ✔ Tailor recognition to the group members.
>
> **Don't**
>
> ⊗ Delay in celebrating the group's successes.
>
> ⊗ "Chew out" a group member for any reason.

on disciplinary procedures. A manager's desire to stop undesirable behavior right away actually undermines the whole premise. There is no way the manager can watch over every employee to ensure safety or quality. A system is needed to reinforce safety and quality. It might be as simple as distributing a bingo chip for every accident-free and defect-free day. As the number of days adds up, so do the number of bingo chips and the likelihood of receiving an award.

Toyota, in Georgetown, Kentucky, provides positive reinforcement for perfect attendance. Each quarter, employees with perfect attendance participate in a drawing for a new Toyota Camry or Avalon.

The nature of positive reinforcement is a little tricky. Most managers believe they already provide positive reinforcement. Yet, if you ask their employees, they don't *feel* it. Daniels finds that most managers make one of the following errors in using positive reinforcement:

Perception errors. Most rewards suffer because they are not tailored to the individual. A perception error involves making assumptions about what other people like. If your perception is wrong, you will think that something will reinforce a person's behavior when, in fact, it won't. For example, not every employee wants more responsibility. Some people don't like public recognition. It's better to ask employees beforehand how they would feel about such rewards.

Contingency errors. Positive reinforcement influences behavior that's occurring at the moment it is given. Managers make contingency errors when they are afraid to interrupt employees while they are working. Instead, they wait until break time to encourage the employees. The result is that the managers end up reinforcing breaks instead of work.

Delay errors. Everyone knows that it is the results that count. The problem is that results might be delayed well beyond the time when the behavior occurred. For this reason, rewarding a delayed result has no effect on shaping future behavior. The reward might even reinforce the wrong behavior. In addition, the reward might be so delayed that it causes anger and resentment.

Frequency errors. The best foundation for involving operators and other employees in a setup-reduction program is building a solid relationship between a manager and his or her employees. Frequently providing positive reinforcement for the desired behavior will increase the frequency of the desired behavior. In addition, if the manager frequently provides positive reinforcement, employees will regard a manager's request for participation in a setup-reduction program as sincere. If the manager approaches employees only when they did something wrong, they will become defensive when the manager asks for their participation in the program.

Involving employees in setup-reduction programs requires antecedents followed by consequences to ensure that reinforcement is:

- Meaningful to the employee
- Earned
- Immediate
- Frequent enough to increase the desired behavior

POINTS TO REMEMBER

☞ Take the time to thank employees for their efforts in reducing setup and changeover time.

☞ Recognize people in a way that is meaningful to them.

☞ Keep it fun!

CHAPTER 10

Every Second Counts

We started Chapter 1 with an analogy about how a race is won or lost in the pits. You don't have to be a race fan to understand that this is true. Even the most casual observer of a NASCAR race can't help but notice that pit stops are a flurry of activity. During a routine pit stop, the pit crew puts on four new tires, fills the gas tank, cleans the windshield, cleans the front grill, slightly adjusts the chassis, and provides fluids for the driver—all in a heart-thumping 20 seconds!

NASCAR is big business. Owners spare no expense in recruiting and training pit crews and supplying them with the tools they need to win a race. Can you think of any reason why we shouldn't give the same degree of attention and importance to setup and changeover practices in our plants?

Now, let's see how the pit crew uses the principles for quick setup and changeover.

Establish the Baseline

The pit crew, also known as the over-the-wall gang, consists of seven members. Each member has a specific job assignment:

1. Jack man
2. Front tire changer
3. Rear tire changer
4. Front tire carrier
5. Rear tire carrier
6. Gas man
7. Catch can man

The jack man is the leader of the group, or *crew chief*. He signals the start and finish of the pit stop. The crew chief and his entire pit crew are fully dedicated to continuously improving their performance because every second counts. Thus, it is not uncommon for the pit crew to videotape and review every pit stop. We can accomplish the same feeling of teamwork and importance in our efforts to reduce setup and changeover time. It's a matter of personal commitment, communication, training, and practice. For a quick refresher on how you can establish a baseline, see Chapter 4.

Separate Internal from External Operations

Just as we do in the plant, the pit crew looks for ways to reduce the total elapsed time of the pit stop by separating operations (see Chapter 5). To make sure that only internal activities are performed while the car is in pit box, the pit crew:

- Presets the jack's lift plate to just below the car's height at the beginning of each race. This saves precious seconds when jacking up the car.

- Puts a quick-tire-change position mark on each side of the car so that the jack man can immediately find the position he needs to be in.

- Arranges the race tires in sets and positions each wheel in relation to where its location will be on the car.

- Fills the gas can ahead of time so that the gas is ready to be poured into the car.

- Prepares the water bottle for the driver ahead of time.

- Places a pit marker on the end of an extension pole to guide the driver into his or her pit box.

Convert Internal into External Operations

As in the plant, whenever possible, the pit crew converts internal activities to external activities (see Chapter 6). For example, they:

- Use studs that are extra long and that have no threads for the first 3/4 inch on each wheel. This saves precious seconds positioning the tire onto the wheel and ensures that the lug nuts aren't cross-threaded.

- Glue the lug nuts onto the wheels before the race. This saves precious seconds when installing the tires and prevents the tire changer from accidentally dropping a lug nut during tire installation.

• Use a special valve on the gas can that fits into the car's port so that the can empties quickly.

Streamline Efforts

After the pit crew has addressed the major causes of slow pit stop times, they examine and streamline the remaining internal and external activities (see Chapter 7). For example, they:

• Store two fuel cans in the far side of the pit box and aim them in the direction of the car.

• Place the catch can next to the fuel cans.

• Use extension poles to clean the windows, clean the front grill, and give the driver the water bottle so that these tasks are performed from behind the wall and out of the way of the other crew members.

• Store the tires at just the right height on the tire stands so that the tire changers can easily carry them over the wall to the car.

• Color code the wrenches so that they are easy to find inside the pit garage.

• Color code the fluid containers to prevent any mistakes when the crew is in a hurry.

• Organize everything inside the pit box and the pit garage to make items easy to find.

Be a Winner

All of these efforts help the pit crew members perform fast pit stops, which can lead to their driver winning the race. You, too, can be a winner in the marketplace race. It is a matter of commitment and practice. Reducing setup and changeover time will make your company better and stronger. It will keep the fun and challenge in manufacturing, too.

POINTS TO REMEMBER

- ☞ Every second counts in racing as it also does in setup and changeover.
- ☞ Establish the same sense of urgency as a pit crew.
- ☞ Separate internal from external activities.
- ☞ Convert internal to external activities.
- ☞ Streamline efforts.
- ☞ Win the marketplace race.

FINAL THOUGHTS

Lean Manufacturing Alone Is Not Enough!

To survive in today's environment, a company needs a two-pronged approach:

- Be the "last man standing" in terms of productivity and cost
- Be the master of product and process innovation and create new business models to respond to and anticipate market needs

CEO Challenge: Adapt or Perish

In the future, a company will need to be

- nimble and agile so that it can quickly respond to threats as well as opportunities;
- innovative in all areas, including
  - business planning;
  - products;
  - manufacturing methods;
  - distribution;
  - customer service.

A company will need to embrace change and learn how to thrive in a constantly evolving environment.

While the company keeps operating, its CEO needs to answer some important questions:

- Are we aware of the competitive threats our firm faces today?
- Are we a "one trick pony?" Will the market we are structured to serve be the same in 3 years?
- Are all employees aware of the environment the company operates in today?
- Do we have a plan as to how we will cope with the internal and external threats and opportunities we face today? Has

this plan been communicated to all employees? Do they know their role?

- What is our plan to be the "last man standing" in terms of productivity and operational excellence?
- What is our plan for future innovation?
- Do our actions convey your commitment to all employees?
- Are we succeeding or losing ground?

THE ROLE OF LEAN MANUFACTURING

☛ Ensure that your firm is the "last man standing."

☛ Ensure that your company remains low-cost/high-quality/innovative cost structure.
  - ◆ Design for manufacture (95 percent of cost in this category)
  - ◆ Office operations
  - ◆ Shop operations
  - ◆ Quality

☛ Ensure that operations run trouble-free so that top management can focus on the future.
  - ◆ Changing markets
  - ◆ External competitive threats
  - ◆ Opportunities

☛ Pilot (CEO) must be able to focus on flying the plane (running the business).

FOR MORE INFORMATION

Quick Setup and Changeover

Productivity Press Development Team, *Quick Changeover for Operators: The SMED System* (New York: Productivity Press, 1999).

Shigeo Shingo, *A Revolution in Manufacturing: The SMED System* (New York: Productivity Press, 1985).

Video Communications, *Manufacturing Insights: Quick Changeover for Lean Manufacturing* videotape (Dearborn, MI: Society of Manufacturing Engineers, 2000).

The Winner's Circle videotape (Tier One Communications: Toronto, Canada).

Reward and Recognition

Aubrey C. Daniels, *Bringing Out the Best in People* (New York: McGraw-Hill, 1999).

INDEX

ABOUT THE AUTHORS

Fletcher Birmingham is president of Summit Business Consulting, Inc., of Hudson, Ohio, providing assistance in strategy, operational excellence, and employee development. His consulting clients include both large and small manufacturing and service organizations.

In addition, Fletcher is director and founder of the Operations and Plant Manager's Association, a network for benchmarking and sharing best manufacturing practices. He is adjunct faculty for the University of Phoenix Cleveland Campus, facilitating graduate courses in operations management, project management, resource optimization, and supply chain management.

Fletcher's work experience spans twenty years in manufacturing and service industries with companies including Hamilton Sundstrand, Lockheed Martin, Scott Fetzer, and Roadway Express.

Jim Jelinek is president & CEO of Moog Flo-Tork in Orrville, Ohio. Moog Flo-Tork is a manufacturer of rotary actuators for military (Navy) and critical service industrial applications. Jim has twenty-eight years of experience in manufacturing.

Additionally, Jim has served as the chairman of the Valve Manufacturers Association and currently serves as the co-chairman of the Submarine Industrial Base Council.